Richard Clarke Cabot

The Serum Diagnosis of Disease

Richard Clarke Cabot

The Serum Diagnosis of Disease

ISBN/EAN: 9783744718769

Printed in Europe, USA, Canada, Australia, Japan

Cover: Foto ©berggeist007 / pixelio.de

More available books at **www.hansebooks.com**

THE SERUM DIAGNOSIS OF DISEASE

BY

RICHARD C. CABOT, M.D.,

Physician to Out-Patients, Massachusetts
General Hospital

NEW YORK
WILLIAM WOOD AND COMPANY
MDCCCXCIX

PREFATORY NOTE.

THIS book is nothing but a compilation. Its aim is to bring together in convenient form the results of the immense amount of work which has been done upon the serum diagnosis since 1896.

My hearty thanks are due to Dr. Henry Jackson, who has kindly allowed me to allude to some of his unpublished cases.

190 MARLBOROUGH STREET, October, 1898.

CONTENTS.

PART I.

	PAGE
INTRODUCTION,	1

CHAPTER I.

GENERAL DESCRIPTION OF THE PHENOMENON OF AGGLUTINATION, . . . 5

CHAPTER II.

TECHNIQUE, ... 10
   I. The Microscopic Test with Fluid Serum, . . . 11
      1. Collecting the Blood, ... 11
      2. Diluting the Blood, ... 12
      3. Examination of Slide and Cover-slip Specimen, . . 13
      4. What Constitutes a Reaction? 13
      5. Precautions and Sources of Error, . . . 15
   II. Time Limit and Dilution, ... 15
   III. Modifications of the Microscopic Method, . . . 17
      (a) Use of Whole Blood, ... 17
      (b) Use of Blister Fluid, ... 18
      (c) Use of Pasteur's Pipette, ... 18
      (d) Centrifugalizing the Blood, ... 18
      (e) Dilution on the Slide, ... 19
      (f) Dilution with Thoma-Zeiss Pipette, . . . 19
      (g) Use of Suspensions, ... 19
      (h) Other Modifications, ... 20
   IV. The Macroscopic or Slow Method, ... 21
      1. With Twenty-four-Hour Culture, . . . 21
      2. With Nascent Cultures, ... 21
      3. Comparative Value of this and the Microscopic Method, . 22
      4. Sources of Error with the Slow Method, . . 23
   V. The Use of Dried Blood, ... 26

CHAPTER III.

   VI. THE CLUMP REACTION WITH DEAD BACILLI, . . . 30
   VII. Measurement of Clumping Power, ... 31

CONTENTS.

CHAPTER IV.

	PAGE
VIII. The Preparation of Cultures,	40
(a) Differences Between Different Races,	40
(b) Effect of Environment on Availability of Cultures for Serum Diagnosis,	43
VII. Preparation of Cholera Cultures,	48
IX. The Clump Reaction with Other Body Fluids,	48
X. The Clump Reaction with Filtered Cultures,	54
XI. "Autoserum Reactions,"	55
XII. Reactions with Chemical Reagents,	55

CHAPTER V.

1. Origin and Nature of the Clumping Power,	56
2. Nature of the Clumping Process,	59

PART II.

CHAPTER VI.

The Occurrence of the Serum Reaction in Typhoid Fever,	61
1. General Statistics,	63
2. How Early Does the Reaction Appear?	64
Summary of Statistics on this Point,	72
3. How Late May Reaction First Appear?	73
4. How Early May Reaction Disappear?	73
5. How Long May Reaction Last?	75
Summary of Statistics on this Point,	77
6. Theory of Late Reactions,	77

CHAPTER VII.

The Serum Reaction in Typhoid Fever (*Continued*),	78
1. Intermittence of the Reaction,	79
2. Relation of the Reaction to Body Temperature,	79
3. The Reaction in Cases Vaccinated against Typhoid,	79
4. Value of the Serum Diagnosis in Typhoid,	79
(a) Inferences from a Negative Reaction,	79
(b) Agglutination in Non-Typhoid Cases,	81
(c) Summary and Interpretation of Results,	87–88

CHAPTER VIII.

Cases Illustrating the Value of Serum Diagnosis,	89–104

CHAPTER IX.

Other Applications of Serum Diagnosis,	105
1. Retrospective Diagnosis of Typhoid,	105
2. The Serum Reaction in the Investigation of Epidemics,	105
3. The Reaction in Infancy and Old Age,	106

CONTENTS. vii

		PAGE
4.	The Reaction in Abortive Cases,	106
5.	The Reaction in Typhoid without Fever,	106
6.	The Reaction in Typhoid without Intestinal Lesions,	107
7.	The Reaction in Typhoid with Double Infection,	108
8.	The Reaction in Cholecystitis and Gall-Stones,	108
9.	The Reaction in Typhoid Meningitis,	109
10.	The Reaction in Melancholic Types of Typhoid,	109
11.	The Reaction as a Means of Distinguishing Typhoid from Allied Infections,	110
	(a) Brill's Cases,	110
	(b) Infection by Gärtner's Bacillus,	110
	(c) "Paracolon" Infections,	111
	(d) "Malta Fever,"	112
	(e) Mountain Fever,	112
	(f) Madagascan Typhomalaria,	112

CHAPTER X.

SERO-PROGNOSIS, ... 113

PART III.

CHAPTER XI.

The Serum Reaction in Other Diseases,	117
1. Glanders,	117
2. Malta Fever,	118
3. Yellow Fever,	119
4. Cholera,	120
5. Bubonic Plague,	124
6. Pneumococcus Infections,	125
7. Colon-Bacillus Infections,	127
8. Tuberculosis,	132
9. Streptococcus Infections,	132
10. Leprosy,	134
11. Diphtheria,	135
12. Tetanus,	136
13. Pleuropneumonia of Cattle,	136
14. Hog Cholera,	137
15. Pictou Cattle Disease,	137
16. Proteus Infections,	137
17. Staphylococcus,	139
18. Anthrax,	139
19. Psittacosis,	139
20. Relapsing Fever,	141
APPENDIX,	143
BIBLIOGRAPHY,	146

THE SERUM DIAGNOSIS OF DISEASE.

INTRODUCTION.

SCOPE AND VALUE OF THE SERUM DIAGNOSIS. ITS USE IN CUBA AND PORTO RICO.

It has been a source of disappointment to many that the study of bacteriology has not furnished to the practitioner as many clews to diagnosis, prognosis, and treatment as might have been hoped, judging from the position of central importance assumed by micro-organisms in modern pathology. Bacteriology has helped our diagnosis of cholera, our diagnosis and prognosis of tuberculosis and malaria, our diagnosis and treatment of diphtheria and tetanus. But of late years, previous to the discovery of the method of serum diagnosis, there has been a decided break in the series of clinically helpful discoveries in the domain of bacteriology. The interest aroused over the problem of acquired immunity has, however, revived our hopes as to the value of bacteriological research, and has led, moreover, to the opening out of several new avenues of approach toward the central mysteries of disease. One of the corollaries or by-products of the immense amount of labor and thought which have been expended of late years upon the problem of immunity is the discovery of the method of serum diagnosis.

I do not know a better way of exemplifying the scope and value of the serum diagnosis than by giving an account of the use which was recently made of it in investigating the fevers so rife this summer among the American soldiers in Cuba and Porto Rico.

In studying the cases of the first batch of soldiers brought from Santiago to the Massachusetts General Hospital it was brought home to all of us who saw them that the ordinary methods of investigation and diagnosis left us very much in the dark. After taking a careful history and making a thorough physical examination we knew very little more than at the start. This was due to several causes.

(1) The soldiers' remembrance of the onset, duration, and course of their symptoms was very imperfect and vague. They often could not remember whether they had had headache, or chills, or not.

(2) Such facts as they could remember were practically the same in every case—both in those which afterward turned out to be typhoid and those that proved to be malaria or dysentery.

(3) Physical examination was surprisingly unproductive. Almost every patient showed emaciation, splenic enlargement, a yellowish-brown tinge to the skin which entirely prevented the recognition of the presence or absence of rose spots. Hardly any of them showed the mental hebetude often seen in typhoid. The stools were not characteristic.

Altogether the cases were very much alike, and we could not wait a week to see whether the temperature curve would clear up the diagnosis. But when the blood was examined we began for the first time to get some solid objective evidence. The typhoids showed (ninety-five per cent of them) a positive serum reaction, the malarias a characteristic parasite, the dysenteries nothing.

This proof of the importance of the serum diagnosis made me very anxious to have the test used in Cuba and Porto Rico, and I accepted the first opportunity that offered and went to Porto Rico on the hospital ship *Bay State*. At Ponce, in the United States General Hospital, I found several hundred fever cases, almost all of whom were receiving large doses of quinine, because the diagnosis between typhoid and malaria had not been clearly established owing to the entire lack of any materials for performing the serum reaction for typhoid. I examined one hundred and seventy-seven cases at this hospital and found not a single case of malaria among them, while ninety per cent of those examined showed prompt and intense serum reaction for typhoid. This evidence, added to that previously obtained by Dr. Leary (pathologist to the hospital) in fifteen cases, was sufficient to check the wholesale administration of quinine and indicate which of the cases needed typhoid diet.

Again, in the diagnosis between typhoid (with jaundice) and yellow fever the test was of value; three out of four cases in one quarantine tent at Ponce—cases suspected of yellow fever—showed a marked and immediate typhoid-serum reaction.

Through the examination of one hundred cases from Utuado and seventeen from Guanica, besides those already mentioned, I was able to convince myself that there was a great deal of typhoid and dysentery but very little malaria among the American soldiers on the island. Out of all those examined only two showed malarial parasites. Among those in Cuba, on the other hand, the malarial fevers predominated greatly over the typhoid. When the Spanish prisoners were kept waiting nearly a week off Portsmouth, N. H., while the physicians attempted to find out whether their fevers were due to typhoid, malaria, or yellow fever, I could not help thinking that a great deal of time could have been saved and much surer results obtained had the method of serum diagnosis been used. As yet, however, the general practitioner has not taken up the method and it remains in the hands of a comparatively small number of men. This is a great pity, since the test is an exceedingly simple one, easily learned and quickly performed, and one needing no extra apparatus provided a good microscope is at hand. The cultures used in the test can be kept indefinitely at room temperature, and no thermostat is needed. The whole process can easily and safely be carried out by the physician in his office without any laboratory facilities and without half the skill or labor necessary to examine urinary sediments. I hope that this book may do something toward making the test more widely used outside of laboratories and in the hands of the general practitioner.

PLATE I.—SERUM DIAGNOSIS—CABOT.

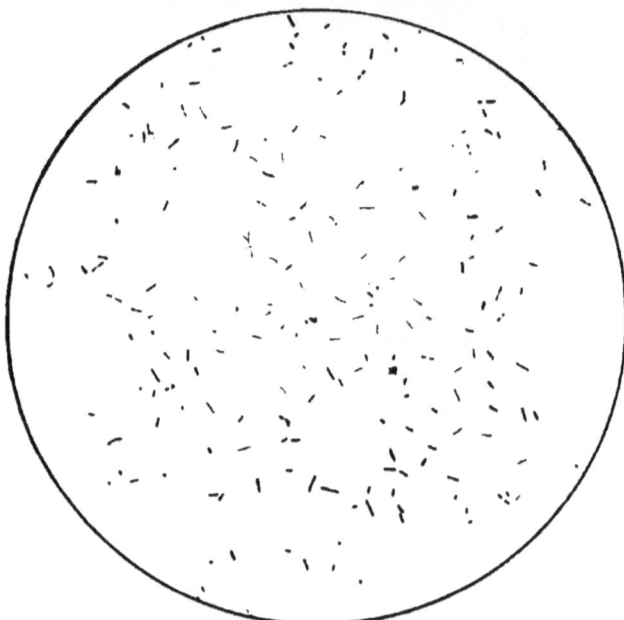

FIG. I.—Bouillon Culture of Typhoid Bacilli before the Addition of Diluted Typhoid Serum. (Magnified 500 diameters.)

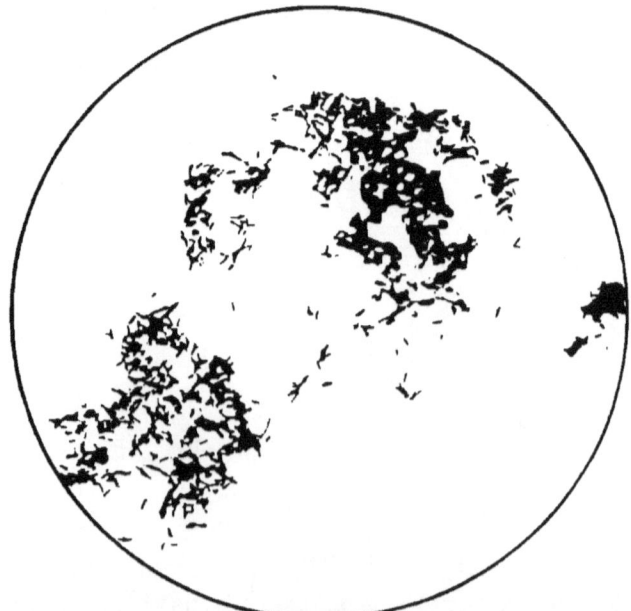

FIG. II.—The Same, Five Minutes after the Addition of Typhoid Serum (dilution 1:10), showing typical clump reaction. (Magnified 400 diameters.)

PART I.

CHAPTER I.

GENERAL DESCRIPTION OF THE PHENOMENON OF AGGLUTINATION.

A DROP of a young bouillon culture of typhoid bacilli, cholera vibrios, or any motile organism examined between a slide and cover-glass with a high-power lens shows the organisms darting about and across the field with great rapidity, and in various directions. There is no tendency for them to adhere to one another. They very often hit against each other, but bounce off and continue their motion at once. (I am speaking now of young cultures under favorable conditions of environment.)

I.

Now if to ten drops of such a bouillon culture, say of cholera vibrios, one adds a drop of blood drawn from an animal strongly immunized against the cholera vibrio, there occurs an agglutination of the vibrios in clumps. This process can be watched under the microscope, in a slide-and-cover-glass preparation. Clumps large enough to be easily visible under a magnifying power of 500 diameters usually appear within a few minutes. The agglutination is accompanied by a loss of motility. The greater the degree of immunity possessed by the animal used, the larger the clumps are and the more swiftly they are produced. If the preparation is preserved for a few hours the clumps may grow so large as to be visible to the naked eye.

The same reaction takes place between typhoid bacilli and the blood of an animal rendered immune to the typhoid bacillus. Any pathogenic motile organism can be used in the same way— e.g., the bacillus pyocyaneus, the plague bacillus, the bacillus coli communis, etc. Again, various of the body fluids other than the blood can be used with similar results; for example, milk, tears, serous or purulent exudates, the juice of various organs, or the dried blood or serum.

II.

Instead of the body fluids of immune animals, the same fluids from patients suffering from any infective disease due to a motile bacillus produce a similar specific clump reaction with the bacillus of the disease in question.

Thus plague bacilli are clumped by the serum of sufferers from the plague, cholera vibrios by the serum of cholera patients, typhoid bacilli by the serum of those in the later weeks of typhoid fever.

The phenomenon above described has been made use of for two purposes:

1. The diagnosis of disease.
2. The identification of micro-organisms.

Given a culture clearly identified as typhoid, one can use it for the identification of cases of typhoid fever when diagnosis is difficult. Or given a case known to be typhoid fever, and one can utilize the blood or serum of the patient as a means of identifying doubtful cultures.

Both of these procedures depend upon the relatively specific nature of the reaction. Typhoid bacilli are not clumped by any serum other than that of a typhoid patient or an animal immunized against typhoid, provided the proper technique is used. Typhoid serum clumps no organism other than the typhoid bacillus.[1]

This description is sufficient to define the general nature of the phenomenon of agglutination, and we may now turn to a short account of the steps by which our present knowledge of the subject has been built up.

HISTORICAL SKETCH.

As long ago as 1889 Charrin and Roger described appearances which we can now see to have been essentially an agglutinative process. They found that the bacillus pyocyaneus, cultivated in the fluid serum of an animal immunized against that bacillus, did not produce any diffuse turbidity of the serum such as appears if the bacillus is grown in the serum of any non-

[1] These last statements are subject to some reservations (see page 15 et seq.).

immune animal. Instead of being scattered through the culture medium, as is ordinarily the case, the organisms "are joined together in little clots, which separate if the test tube is shaken, but sink to the bottom again shortly. . . . Microscopically the microbes are joined in chains, and show a great tendency to group themselves together instead of swimming about freely as normal bacilli do. *On les voit s'enchevêtrer en petits amas, ce qui explique sans doute l'aspect grumeleux des cultures.*" [1]

Here, in 1889, was a clear recognition of the phenomenon; yet this observation did not seem to be recognized as important at the time, either by the observers or by any one else. It slept unnoticed and forgotten until after Widal's application of the principle here involved to the diagnosis of disease. Then on ransacking the literature this long-forgotten observation was brought to light. It had been thought of solely from the point of view of a study of immunity, and no practical application to diagnosis had occurred to any one till 1896.

Metchnikoff had indeed, in 1891,[2] established analogous facts concerning the vibrio bearing his name, and the diplococcus lanceolatus. He found that when grown in the serum of animals immunized against them, these micro-organisms lost their motility and formed "des paquets plus ou moins grands." Metchnikoff was evidently struck by this fact. "*Ce fait*," he says, "*présentant une importance générale, doit être examiné de plus près.*"

Evidently he had something like a presentiment of the future importance of these facts; yet he never followed up the clew thus obtained, because, working with a bacillus which he supposed (wrongly) to be that of hog cholera, he found no reaction similar to that established for the pneumococcus and the vibrio Metchnikovi. In point of fact the hog-cholera bacillus *is* agglutinated by immune hog-cholera serum, as has been since proved; but Metchnikoff had lost his clew, and so dropped his investigations along this line. Two years later, in 1893, Issaeff confirmed Metchnikoff's observations on the pneumococcus; but still the facts attracted no general attention, and their discoverers did not feel encouraged to go on with the subject.

[1] "Note sur le développement des microbes pathogènes dans le sérum des animaux vaccinés." Comptes rendus de l'Académie des Sciences, November 9th, 1889.
[2] Annales de l'Institut Pasteur, 1891, p. 473.

In 1894 Pfeiffer made practically a fresh start from a slightly different point of view. Hitherto the effects of immune serum on the growth of bacilli had been studied outside the body in test tubes. Pfeiffer experimented to determine the behavior of cholera vibrios in the peritoneal cavity of guinea-pigs immunized against this micro-organism. He showed that in the peritoneal cavity of normal guinea-pigs the vibrios multiplied with great rapidity and suffered no considerable modification in their morphology, while in the immunized guinea-pigs they were quickly immobilized and broken up into small granules.

This reaction of the body fluids of an immunized animal is known as "*Pfeiffer's phenomenon.*" Pfeiffer's chief interest in it was due to the light it threw upon the question of immunity and the assistance afforded by it in the identification of the cholera vibrio. According to Pfeiffer the phenomenon is a strictly "*specific*" one, that is, it does not occur when any organism other than the cholera vibrio is put into the guinea-pig immunized for cholera, nor when the cholera vibrio is put into normal pigs or those immunized against other micro-organisms.

Pfeiffer's later studies showed that the same reaction takes place, if some serum from an immunized animal be put into the belly of a normal guinea-pig along with the vibrios. The latter are reduced to motionless granules, as in the body of the immunized animal.

In 1895 Bordet took the next step by discarding the use of the guinea-pig's peritoneum, and mixing the immune serum with the vibrios in a test tube. Here the reaction occurred as before, though somewhat less markedly.

In 1896 Pfeiffer extended the application of his method to the recognition of typhoid bacilli. Pfeiffer's reaction, as will be noticed, is distinctively different from the agglutinative reaction. Pfeiffer noted no special clumping, and called attention chiefly to the breaking down of the bodies of the bacilli into granules and their final disappearance.

The first systematic and thorough studies of the agglutinative reaction were published by Durham, who communicated the results of his investigations in Grüber's laboratory at a meeting of the Royal Society, January 3d, 1896. "In this and the rapidly following papers of Grüber and Durham the real importance and general character of this reaction *with immune serum* were for the first time made clear. The macroscopic and microscopic tests, the importance of dilutions, quantitative estimations of agglutinative power, the value of the differentiation of bacterial species, the determination of a previous attack of typhoid fever, and many other details, were described."[1] Hence there are some writers who prefer to call the agglutinative reaction the "Grüber-Durham" test.

But if we are to go back to the first description of the reaction, the names of Charrin and Roger should be those connected with the test. If, on the other hand, we are to give the credit to him who first *applied* the reaction to the diagnosis of disease, there can be no doubt that it is to Widal that the merit belongs. For Widal was the first to point out that the agglutinative reaction is present not only in the serum of artificially immunized animals, and in the blood of convalescents, but during the *period of infection*, in the early stages and at the height of the disease. Widal has steadily insisted that the reaction is not one of immunity, but of infection, and his contention, at least on the negative side, has been entirely borne out by later investigation. Widal's first paper was read June 26th, 1896, and at once attracted attention all over the world. If any observer's name is to be attached to the reaction, it should certainly, in my belief, be that of Widal. The word "serodiagnostic" was invented by him.

Widal's discovery was rapidly confirmed by many French writers, among whom Dieulafoy, Achard, Chantemesse, Siredey, Menetrier, Nicolle and Halpré, and Courmont were the earliest. America was the next country to wake up to the importance of the discovery, Wyatt Johnston beginning his researches in September, 1896, and Miller at the same time.

Germany was very slow to take up the new reaction, and it was not until after the method of serum-diagnosis was well established in France and in this country that the earliest German articles began to appear.

For further details of the extension and modification of the reaction, the reader is referred to the body of the work.

[1] Welch: Journal of the American Medical Association. August 14th, 1897.

CHAPTER II.

TECHNIQUE.

The various methods of demonstrating the clump reaction fall naturally into three groups:
1. The microscopical or quick test of the fluid serum or blood.
2. The microscopical or quick test of the dried blood.
3. The macroscopic or slow test.

Each of these has certain advantages and disadvantages, and each has been subjected to various modifications in different hands. Each has its advocates, and many claim that only one of the methods is accurate—the others being unreliable. The most important thing, however, as has been pointed out by a committee of the American Medical Association, of which I was a member (see Appendix), is that each observer should select some one method and stick to it, so that by experience in a large number of cases he may become thoroughly familiar with its working—with the more frequent sources of error and the ways of avoiding them, with the degree of its accuracy, and the proper method of drawing conclusions from its data.

A method by which A gets the best results is not necessarily best for B, and the time which must be expended in making extensive comparative tests of the different methods had better be used in acquiring that skill in the application of some *one* method which comes only by long familiarity and practice. This applies to many other procedures in microscopical technique. To learn one method of examining sputa or blood films, and get thoroughly familiar with that, is the safest procedure for most of us. The greater number will in any case be best suited by some single one of the standard methods, but there will always be champions of other methods. Thus, while the majority of observers prefer the first of the three methods above mentioned, there is no question that in the hands of Wyatt Johnston the second method is as reliable and delicate as any, his enormous experience being chiefly with that, while Wright and some English observers get excellent results by the third.

TECHNIQUE.

I. The Quick or Microscopic Test with Fluid Serum.

1. *Collecting the Blood.*

The blood is best obtained from the ear as in other hæmatological work, and the puncture is made in the ordinary way with a bayonet-pointed surgical needle (Fig. 1) or a fine-pointed

Fig. 1.

knife. I have used mostly a narrow-bladed knife, but when this was not at hand I have found no difficulty in getting enough blood from a needle puncture. It is not necessary to clean either the needle or the skin. In over ten thousand such punctures made without any antiseptic precaution I have never seen the slightest sign of sepsis. After puncturing the most dependent portion of the lobule of the ear, milk it strongly downward toward the point of puncture. The blood is thus forced out in large drops. To collect the blood I have used for the most part small test tubes (Fig. 2). Sometimes the blood will fall freely from the ear into the tube. In other cases we have to put the edge of the tube against the ear and scrape off each drop as it emerges so that it runs down the inside of the tube and collects at the bottom. In this way about fifteen large drops of blood can be obtained one by one by milking the ear. This will furnish two or three large drops of serum, which is enough for clinical work.

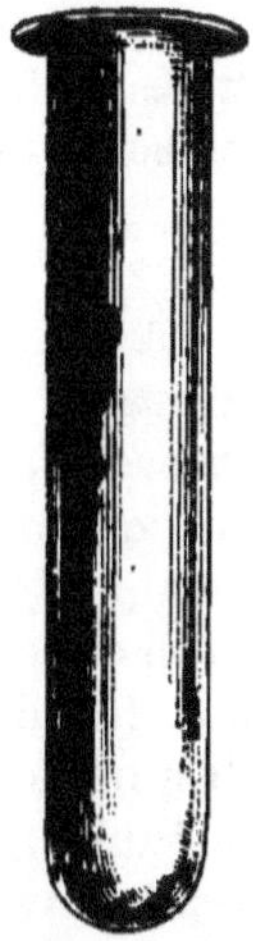

If we need a larger quantity of blood for purposes of investigation and experiment, we have only to continue the milking process or prick the other ear. It is unnecessary to puncture a vein, as the Germans do.

The process just described gives abundant opportunities for contaminating the blood; but this is no harm, since it has been shown that even a considerable degree of putrefaction does not interfere with the reaction. The test tubes, therefore, need not be sterile, but we must take care that they contain no traces of antiseptics or of blood from a previous case of the disease

Fig. 2.—Test Tube (actual size).

whose presence or absence we are to determine. For example, if a tube containing traces of typhoid blood be used to test the blood of another case suspected of being typhoid, a positive clump reaction may be obtained, and yet the case may turn out not to be typhoid, the clumping being caused by the agglutinating material of the previous case. I have known mistakes to occur in this way; but to guard against it, one does not need to sterilize one's tubes, but only to have them reasonably clean.

The first time that I tried this test, following the description given in Widal's first communication on the subject, I waited all one afternoon for the blood to coagulate in the tube so that the serum would exude upon the surface. To my chagrin this did not occur, and it took me some time to discover that the serum was there all right, only pinned down beneath the clot which often adheres tightly to the sides of the tube. It is necessary, therefore, as soon as clotting occurs (which is usually within two minutes), to separate the clot from the sides of the tube, using a bit of wire or the blade of a penknife. The serum then exudes and is ready for use.

It is an advantage to have a few red corpuscles in the serum, as they give us something to focus on when the mixture of serum and culture is under the microscope. By daylight it is sometimes rather difficult to focus on the colorless bacilli, so that the presence of corpuscles is not to be deprecated.

2. *Dilution*.

A drop of the serum so obtained is now to be added to forty drops of a bouillon culture or emulsion of the bacilli of the disease for which we are testing. [The method of preparing the cultures is explained on page 40.] If we use the same dropper to measure out the forty drops of culture and the one drop of serum, the drops will be approximately of the same size and the dilution sufficiently accurate. An ordinary medicine-dropper answers very well. The mixing can be done in a small test tube like that used to collect the blood, or in any other convenient vessel. The blood serum and culture should be well stirred and mixed, and is then ready for examination.

Professional bacteriologists generally recommend the use of a hanging-drop preparation, but this is wholly unnecessary, since it is just as well and quicker to use an ordinary slide and coverglass, as was sensibly suggested by Widal in his early communications. Bacteriological technique is complicated enough at the best, and all unnecessary manipulations and apparatus should be avoided.

A drop of the mixture of culture and serum is therefore put upon a slide, covered with a cover-glass and examined at once. If no reaction has taken place when the *time limit* (see below page 15) has expired, another drop of serum should be added to the mixture of serum and culture, thus giving a 1 : 20 dilution. If on examination the reaction is still absent, two more drops of serum should be added so that the mixture is 1 : 10. If no reaction occurs at this dilution we should consider the test negative. Instead of diluting the same bulk of culture we can prepare three tubes, one containing ten drops, one twenty, and one forty. To each of these a drop of serum can be added and the test carried out in each as above described.

3. *Examination of Slide and Cover-Slip Specimens.*

A dry lens of high power, such as a No. 7 Leitz, is preferable to an oil-immersion lens. The bacilli can be easily seen with this power, provided a No. 3 or 4 eyepiece is used, and the light is better suited for the reaction than with the immersion lens. Moreover, the oil of the immersion lens may cause the cover-glass to stick to the objective rather than to the slide, and focussing is thus made difficult. Even with a No. 5 objective of Leitz I have had no difficulty in making out the bacilli.

Artificial light of any kind—gas, electricity, or oil—is preferable to daylight. If daylight is used a small aperture diaphragm works best.

The slide and cover should be carefully cleansed so that no traces of antiseptics or of agglutinating substance left over from a previous case can lead to mistakes.

4. *What Constitutes a Reaction?*

In some cases the instant we look at the preparation under the microscope we recognize the presence of large clumps of bacilli isolated from each other and motionless. In the interspaces between the clumps there are no motile bacilli. All is quiet and apparently lifeless, the sharpest possible contrast to the incessant, furious activity of the bacilli in pure bouillon cultures.¹ This is the so-called "complete" or "typical" reaction. In such cases it is instructive to prepare a slide and cover-glass specimen of the *pure* culture without any serum, and then to allow a drop of the typhoid serum to flow in under the cover-glass while its effect upon the bacilli is watched through the microscope. Clumping may occur instantaneously, or gradually in the course of minutes or hours. The heaps of bacilli usually contain no foreign matter, although at times a bit of fibrin may appear to serve as a nucleus for a clump.

In very marked reactions a mottling of the cover-glass preparation can be seen even with the naked eye, owing to the great size of the clumps formed (Greene: New York *Medical Record*, December 5th, 1896).

Such a typical reaction is, however, rare. In most cases there are some isolated bacilli outside the heaps, and among these isolated organisms as well as at the margin of the clumps there is still a certain amount of motion. A peculiar motion seen under these circumstances is a spinning of the bacillus round one of its own ends as an axis. Durham first called attention to this. It is never seen in isolated bacilli, but always in those at the margin of a clump, and one cannot help thinking that the organism is swinging at the end of one of its own flagella, the latter being entangled in the clump.

When the reaction is feeble or when it is just about to appear one often notices what Widal calls "agglutination centres" ("centres agglutinatifs"). The bacilli are still more or less motile and nowhere tightly clumped, but appear to be drawing together in loose groups, not at first touching each other at all. Sometimes they are arranged a little like the spokes of a wheel. The bacilli appear to be drawn toward the imaginary centre of the group, and often look as if they were trying to resist the clumping influence and to separate themselves.

Clumping and slowing or cessation of motion are the essentials of the reaction, *provided* they take place: (1) In a culture containing no clumps previous to the addition of the serum; (2) within a certain time (the limit being variously fixed by different observers) and despite a certain degree of dilution of the serum.

---

¹ The addition of non-typhoid blood to a pure culture often accelerates the motions of the bacilli.

PLATE II.—SERUM DIAGNOSIS—CABOT.

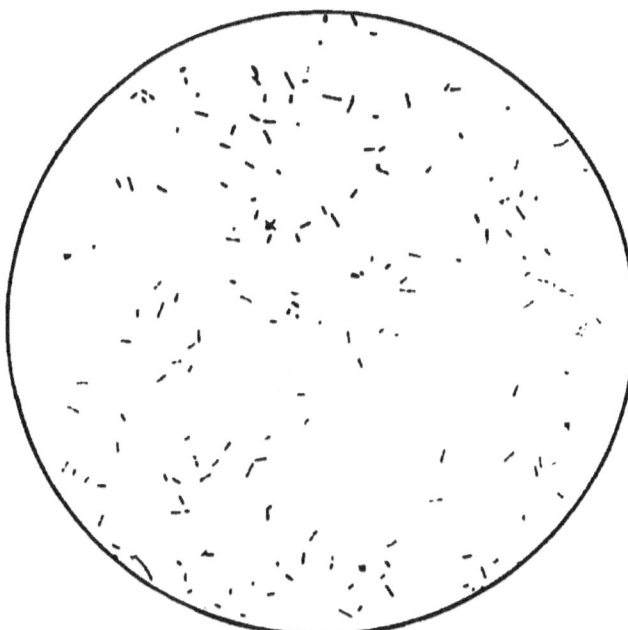

Fig. III.—Earliest Stages of Clumping (Widal's "Agglutinative Centres"). (Magnified 500 diameters.)

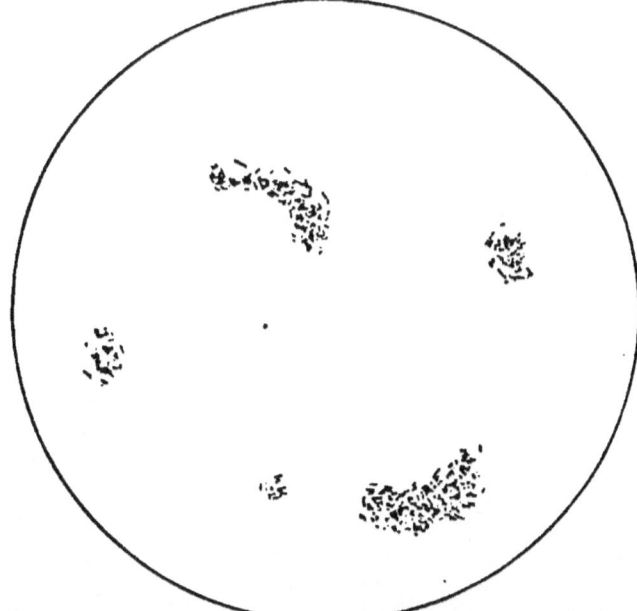

Fig. IV.—"False Clumps," due to Imperfect Emulsion of a Solid Agar Culture. (Magnified 500 diameters.)

PRECAUTIONS AND SOURCES OF ERROR.

1. *Preliminary Examination of the Pure Culture.*

It is a necessary precaution, but one often forgotten, to examine a drop of the pure culture before the addition of any serum, so as to make quite sure that no clumping has taken place. Certain cultures, particularly those containing a sediment or a pellicle, or very old cultures, contain clumps identical in appearance with those produced by typhoid serum. A drop of every culture used should therefore be examined between slide and cover-glass immediately before any serum is added. If any clumps are present it is not safe to use the culture at all.

The bacilli should be isolated and actively motile. [These points will be more fully discussed when we come to speak of the preparation of cultures, see pages 40 to 48.]

2. *Time Limit and Dilution.*

Even after we have made ourselves certain that our culture contains no clumps before the addition of serum, and are sure that the clumps are due to the action of the serum on the culture, even then we are not sure that clumping means a positive reaction. Only when clumping occurs *within a certain time and in a certain degree of dilution* is it of diagnostic importance. These points were not sufficiently emphasized in the earliest communications on the subject. Widal paid little attention to the time limit, and other observers following Johnston's method (as they understood it) were negligent in the matter of dilution.

Now it has been established beyond any doubt that in cultures prepared in the ordinary way, clumping can be produced by normal serum if undiluted and left in contact with the bacilli for a sufficient length of time. In other words, the test is *quantitative* and not *qualitative*. Typhoid serum differs from normal serum in that it will clump typhoid bacilli in a shorter time and in greater dilution than will normal serum or that of other diseases. The same is true of cholera serum in its relation to cholera vibrios, and presumably with the sera of all diseases in which the agglutinative reaction has been found.

The degree of dilution and the time limit cannot be considered separately. A dilution which is correct with one time limit will not do with a longer time limit. For instance, Widal's original dilution figure, 1:10, is perfectly satisfactory in typhoid fever if our time limit is fifteen minutes. That is to say, any clumping of typhoid bacilli taking place within fifteen minutes in a dilution of one part of serum to ten of culture is to be considered a positive typhoid reaction. But if with this same dilution we allow a time limit of one or two hours, clumping not infrequently occurs with sera other than typhoid. Hence the reproach frequently brought against Widal and others who have been content with the 1:10 dilution. German observers generally, especially Stern, Fraenkel, Kühnau, and Haedke, have maintained that the dilution should be 1:25, 1:40, or even 1:50. But these observers have allowed the serum and culture to be in contact for from one to two hours.

One to forty is the proper dilution with a time limit of one hour, 1:10 is equally reliable with a fifteen-minute limit. Discrepancies in the technique of different observers in regard to these points account for such contradictory reports as have been occasionally heard amid the general and remarkable unanimity in regard to the serum reaction.

Kühnau (*Berliner klinische Wochenschrift*, No. 19, 1897) has urged the importance of higher dilutions than those usually employed. He bases this appeal upon results obtained with the long time limit generally employed in Germany—two hours or more. As I have already pointed out, the question of the degree of dilution that should be used in serum diagnosis cannot be separated from the question of time limit.

A dilution of 1:10 which Kühnau condemns is perfectly satisfactory when used with a fifteen-minute time limit. With the longer time limit employed by Kühnau the 1:50 dilution which he recommends is certainly advisable. He tested 50 cases without any sign or history of typhoid; clumping occurred in 8 at from 1:10 to 1:20, 4 at 1:30, 3 at 1:35, and 1 at 1:50.

In accordance with this and other similar observations the 1:10 dilution has been very generally (and I think unnecessarily) abandoned outside of France.

For example: Durham uses a dilution of 1:17 or 1:20. Scheffer uses a dilution of 1:20 (slow or macroscopic method). De Rochemont and Fraenkel use a dilution of 1:25 (slow or macroscopic method). Kolle a dilution of 1:30 (quick method). Grünbaum and Grüber a dilution of 1:32 (quick method with one hour time limit). Stern and Kühnau a dilution of 1:50 (after two hours in a moist chamber on a thermostat). Wilson and Westbrook a dilution of 1:50, with a two-hour time limit.

Park considers that in all positive cases a slowing of motion and some tendency to clumping appear *immediately* in a 1:10 dilution. The reaction is always to be considered complete in half an hour. My own experience coincides entirely with this, but I think fifteen minutes may be allowed for the completion of the reaction. With the higher dilutions there may be very little change within this period.

Modifications of the "Quick" or Microscopic Method.

1. *Use of Whole Blood.*

Instead of serum one can use the *whole blood* in a fluid state. The use of the whole blood dried will be described later (see page 26). Instead of waiting for the serum to separate from the clot we may draw the drop of blood directly into the ten drops of culture previously measured out in a test tube. This is a very quick and simple method successfully employed by Widal, McWeeney, Delepine, Phühl, Coleman, and others. I have used it in over eight hundred cases without perceiving any drawbacks.

Ten drops of the carefully tested culture are measured out with a medicine-dropper into a two-inch test tube. The medicine-dropper is then washed and dried and carried with the test tube of culture to the patient's bedside. In private practice or when the blood has to be carried some distance the tube may be corked to prevent spilling. In hospital work I have generally carried a number of test tubes (each containing ten drops of culture) from the laboratory to the wards and back in an ordinary test-tube rack.

There is no danger in such a procedure if we are careful not to spill. Arrived at the bedside we draw some blood in the ordinary way, suck it into the medicine-dropper, and expel one drop into the ten drops of culture in one of the test tubes. If the same dropper is used for measuring first the culture and subsequently the blood, the dilution is sufficiently accurate. In this way one can go from bed to bed in a hospital ward and do fifteen tests in as many minutes. The bulk of the red corpuscles rapidly settles to the bottom of the tube, and such as remain in the supernatant fluid do not in any way interfere with the reaction, while on the other hand they make the focussing process quicker and easier. If one takes a microscope with him to the bedside the test can be finished up at once. There is among many a superstitious horror of doing bacteriological work in a hospital ward, but with ordinary common sense and care there is no danger in it and much time is saved.

2. *The Use of Blister Fluid.*

Pugliesi, Biggs and Park, and some other observers have used the serum from a blister instead of drawing blood. A cantharides blister the size of a five-cent piece will form in from six to eighteen hours. The serum so obtained is admirably clear, and by some preferred to the serum separated from clotted blood.

3. *Use of Pasteur's Pipette.*

Various instruments have been devised for collecting serum aseptically. The blood can be sucked into a syringe from a puncture of a vein at the bend of the elbow and then expelled into a sterilized test tube. For transportation it is convenient to suck the serum into a modified Pasteur's pipette (Fig. 3) which is then sealed by heat at the extremities. Serum so collected will keep for an indefinite period and can be transported across the ocean. It seems to be unnecessary, however, to be careful as to asepsis when we are using the quick method, since Bensaude and others have found that serum preserves its agglutinating power even when putrefaction is far advanced. It is only when we are using the slow or macroscopic method that we need to be careful as to cleanliness.

4. *Centrifugalizing the Blood.*

The blood when collected may be at once centrifugalized and the plasma separated in this way before any clotting takes place. Stern mixes the fresh drop with the culture (1:10) and then centrifugalizes the mixture. In typhoid cases he often is able to see macroscopically a greater transparency in the fluid than in normal cases, the clumps going to the periphery with the corpuscles and leaving the bouillon clear, while separate and motile bacilli are unaffected.

5. Dilution on the Slide.

Some observers prefer to mix the serum and culture directly on the slide to be used in examining it. For instance, McWeeney gives the following technique: "A perfectly clear slide and cover-glass having been made ready, one end of the bulb (see Fig. 3) containing the blood is broken off, and the other or closed end is brought near a small Bunsen flame, the bulb being carefully held with forceps in the right hand. Expansion of the air in the closed end of the bulb causes a drop of serum to exude from the broken-off end, and this should be carefully caught on the middle of a slide, which is held in the operator's left hand. A number, usually nine, of similar sized drops of culture are then taken up in succession with a platinum loop of suitable size, and placed so as to form a circle of drops around the central one composed of serum. With a platinum loop all are now well mixed, the cover-glass is at once applied, and the preparation observed under the high dry objective."

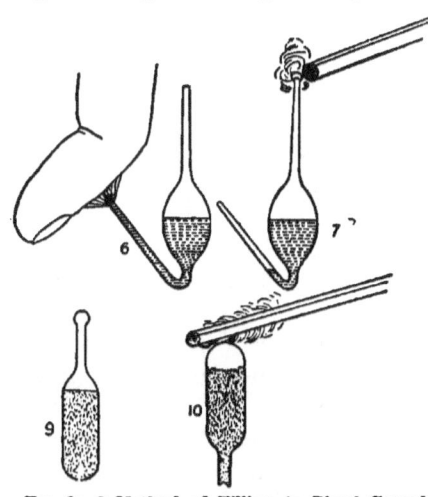

FIG. 3.—6, Method of Filling in Blood Capsule from Finger; 7, showing method of sealing upper end of blood capsule; 8, glass capsule filled in with dead bacterial culture; 9, method of emptying dead bacterial cultivation out of capsule. See page 26 on use of dead cultures.

6. Dilution with Thoma-Zeiss Pipette.

Some are not satisfied with the measurement by drops described above, and desire more accurate dilutions. The measuring pipette of the Thoma-Zeiss hæmocytometer can be used for this purpose, or the special apparatus of Grünbaum or Wright may be employed.

7. Use of Suspensions.

If one has no young bouillon cultures at hand it is sometimes convenient to use a *suspension* of dry agar or serum culture in sterile bouillon. A loopful of the dry culture is rubbed against the inside of the bouillon tube until the bacilli are thoroughly dissociated. Some English observers prefer the use of suspensions to that of cultures even when the latter are available. The only objections to suspensions are (a) the time consumed in making them, and (b) the difficulty of thoroughly separating the bacilli from one another. A drop of the suspension must be examined before the addition of any serum to ascertain that no "preformed" clumps are present. I cannot help thinking that some of Le Fevre's positive reactions in non-typhoid cases are due to his use of suspensions.

8. *Other Modifications.*

Haedke (*Deutsche medicinische Wochenschrift*, January 7th, 1897) reports that mixing a loop of solid culture with typhoid serum on a glass slide sometimes gives very beautiful reactions instantaneously. C. L. Greene[1] has used this method in 150 cases with entire success, but Haedke does not consider it entirely reliable. He often uses the water of condensation from an agar tube instead of bouillon in carrying out the twenty-four-hour or macroscopic method (*vide infra*), which he prefers.

Stokes (New York *Medical Record*, January 9th, 1897) advises that all specimens be stained and preserved for reference. There is no difficulty in staining the bacilli in clumps. A drop of the serum and culture mixture is simply dried on a cover-glass and stained with any of the ordinary aniline dyes.

McWeeney (*Lancet*, May 14th, 1898) notes that if a hanging-drop mixture of bouillon culture with one per cent. serum (*i.e.*, 1:100) be left four hours at 37° C., the bacilli remain separate and motile if the serum is not from a typhoid case; but grow into beautifully twisted and convoluted but motionless *chains* in case the serum is from a typhoid-fever patient.

McFarland (New York *Medical Journal*, September 25th, 1897) makes capillary tubes of a certain size by drawing out ordinary glass tubing. Tubes of the same size must take up the same weight of blood, and thus an absolutely accurate dilution can be made. Ten, fifty, or one hundred times the weight of blood used is measured out in a small crucible into which the filled capillary is put and then crushed so as to set free the whole of its contents.

I have no doubt that any one of these methods is substan-

[1] International Clinics, 1898, p. 168.

tially accurate, provided one gets used to it by experience with its various details in many cases and learns how to avoid the special sources of error connected with it.

I believe myself that most errors arise either:

(a) From the presence of preformed clumps in the culture prior to the addition of serum.

(b) From neglect of the proper rules for dilution and time limit (see page 15).

(c) From ignorance of just what consitutes a positive reaction with the special technique in use (see page 10).

II. THE SLOW OR MACROSCOPIC METHOD.

Widal's first communication described a method of performing the tests quite different from that just described. This method, though now largely abandoned in favor of the quick or microscopic method, is still adhered to by some few German and English observers,[1] and has certain advantages for those who do not own a microscope.

The macroscopic test can be performed in two ways:
1. With a twenty-four-hour culture.
2. With a "nascent" culture.

1. *With a Twenty-four-hour Culture.*

One part of blood or serum *collected under antiseptic precautions* is added to ten or more parts of a twenty-four-hour culture of the bacilli of the disease under investigation and the mixture is left in the thermostat at blood heat. With a serum of powerful agglutinative properties the culture begins to lose its diffuse turbidity within a few hours. The bouillon is at first filled with coarse flocculi, which later settle to the bottom of the tube, leaving the supernatant liquid clear. With less potent serum the flocculi may not attain a sufficient size to make them sink to the bottom at all, or the process may be prolonged to twenty-four hours or more before completion.

2. *With a Nascent Culture.*

Instead of using a full-grown culture in the manner just described, we mix the suspected serum with ten or more parts of sterile bouillon and then add to this mixture a little dry or

[1] *E.g.*, Breuer, Haedke, de Rochemont, Wright, Durham.

fluid culture of bacilli. The growth of the bacilli in the bouillon is then more or less retarded by the influence of the serum according to the degree of its potency. In from four to seven hours flocculi appear, and in from twelve to twenty-four hours the bottom of the tube is covered with a precipitate of gray masses, leaving the bouillon clear. Shaking the tube will not totally dissociate these masses.

Sometimes a precipitate will form within a few hours and then be quickly diffused again, the bacilli regaining their liberty. A control tube containing normal serum culture should be placed in the thermostat, together with the tube containing the suspected serum.

Doubtful reactions are not infrequent. It is then necessary to have recourse to the microscope to settle the question. Pick states that the macroscopic test can be carried out without a thermostat. The reaction occurs in positive cases in from eight to sixteen hours at room temperature. C. L. Greene has observed a macroscopic mottling even in slide-and-cover specimens prepared by his method, due to the large-sized clumps.

Comparative Value of the Quick and the Slow Methods.

The blood, as has been said, must be collected aseptically if the slow method is to be used. This means that we must puncture a vein. Bensaude attaches a bit of sterilized rubber tubing to the large end of the needle of an ordinary subcutaneous syringe also sterile. When the needle has been introduced into the vein the blood flows through the tube and is allowed to drop from this into a sterilized test tube. One must be careful to close the free end of the rubber tube until the blood begins to flow, so as to guard against the entrance of air into the vein.

The labor of puncturing a vein with all the attendant precautions will always prevent the slow method from being generally adopted. Its slowness and the need of puncturing a vein are not compensated for by its supposed greater reliability. Our judgments in the cases that react positively by this method are pretty sure to be right, but many that would be negative or doubtful by this method would be rightly recognized as positive under the microscope.

Another difficulty with the method is that the formation of

flocculi may take place, and then cease within a short time, so that, unless the preparation is constantly watched, one is liable to miss seeing the reaction altogether. This is a great annoyance. Its necessity, however, is admitted even by its advocates (*e.g.*, Haedke). Haedke thinks some change in the appearance of the bouillon is generally to be seen within six or eight hours. He places a good deal of reliance on the effects of shaking the tube as a means of excluding pseudo-reactions. In a typhoid case shaking will not permanently restore cloudiness to a tube that has begun to look flocculent. Pseudo-reactions occurring in diseases other than typhoid (*e.g.*, in tonsillitis) can be proved such by observing that shaking *permanently restores* the diffuse turbidity to the bouillon. If any contamination with other organisms has taken place, the reaction may be entirely masked by their growth, unless recourse is had to the microscope. There one can often recognize the lumps separate from the contaminating bacteria.

With the slow method some errors are avoided, but many diagnoses are impossible which could easily be made by the "quick method." Despite its complication, its slowness, the thermostat method may occasionally be of value as a control. Such at any rate is the opinion of de Rochemont (*Münchener med. Woch.*, 1897, No. 5) and others, and Wright seems to prefer it.

Sources of Error with the Slow Method.

Among the sources of error in the macroscopic method Sabrazés and Hugon (*La Semaine Médicale*, 1897, p. 13) mention the following:

Clumping may occur with non-typhoid serum:

(*a*) If the culture contains a precipitate around which the bacilli can assemble themselves.

(*b*) If when using the method of "nascent culture" one puts in too large bits of solid culture or fails to dissociate the bacilli thoroughly.

(*c*) If the platinum wire used for the transfer is too hot.

(*d*) If the culture or the serum is contaminated.

(*e*) If too large an amount of serum is added when performing the test with "nascent bacilli." Even a normal serum in large quantities may prevent altogether the multiplication of the bacilli. Then they fall to the bottom, leaving the supernatant fluid clear as in the positive reaction for which it might easily be mistaken.

The same observers note that by preparing stained specimens from the precipitated clumps in positive cases, a considerable amount of granular degeneration is to be observed in the bacilli composing them.

Stern (*Berliner klinische Wochenschrift*, 1897, Nos. 11 and 12) considers the twenty-four-hour method unreliable because (according to him) the formation of a visible precipitate depends not only on the agglutinative power of the serum, but upon its bactericidal action, and upon the quality of the bouillon. Though less sensitive than the microscopic test, it cannot be depended on (Stern thinks) as a control of the latter, since he obtained a visible precipitate in two non-typhoid cases even with the twenty-four-hour method and with a dilution of 1:20.

Vanlair and Beco (*La Semaine Médicale*, 1897, p. 15) experimented upon sixteen cases of (clinically) sure typhoid, using both the macroscopic and the microscopic method in each. Five out of the sixteen showed clumps *only microscopically*, the tubes showing no visible precipitate after twenty-four hours in the thermostat. One of these five was proved by autopsy to be typhoid. On the other hand, these observers have found the microscopic method *too sensitive*, in that it sometimes shows some clumping with non-typhoid blood. They conclude, therefore, that no case can be positively identified as typhoid unless the blood responds to *both* tests, macroscopic as well as microscopic.

Wright (*British Medical Journal*, February 5th, 1898, and earlier) has suggested the use of "*sedimentation tubes*" for carrying out the macroscopic test. These tubes have two bulbs (see Fig. 4) and are used like a Thoma-Zeiss blood-pipette, so that dilutions of 1:10 can easily be made. The upper bulb is heated. As it cools, the pointed end of the pipette is dipped into the serum to be tested, which is at once forced into the tube by atmospheric pressure. When the column has entered the tube a short distance the point is marked, the tube is drawn out of the serum and a bubble of air is allowed to enter, after which the point of the instrument is immersed in the culture or emulsion to be used until the column has risen to the point marked. This process is repeated as many times as we desire to dilute the serum; *e.g.*, ten times for a 1:10 dilution.

Or we may dilute five times with normal salt solution, and then add an equal volume of culture (see cut). The lower end of the

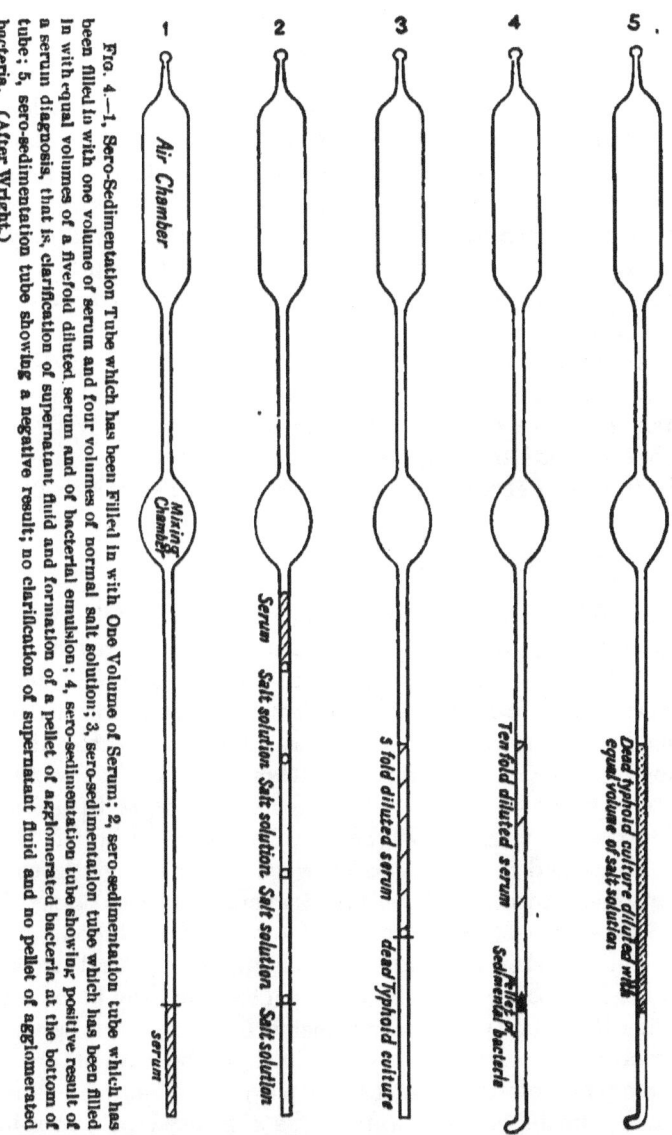

Fig. 4.—1, Sero-Sedimentation Tube which has been Filled in with One Volume of Serum; 2, sero-sedimentation tube which has been filled in with one volume of serum and four volumes of normal salt solution; 3, sero-sedimentation tube which has been filled in with equal volumes of a fivefold diluted serum and of bacterial emulsion; 4, sero-sedimentation tube showing positive result of a serum diagnosis, that is, clarification of supernatant fluid and formation of a pellet of agglomerated bacteria at the bottom of tube; 5, sero-sedimentation tube showing a negative result; no clarification of supernatant fluid and no pellet of agglomerated bacteria. (After Wright.)

pipette is then sealed and the instrument placed in an upright position and left for twenty-four hours.

If the case is one of typhoid, the bacilli settle in a mass to the lower end of the tube. With non-typhoid blood the mixture remains turbid. (See Fig. 4.)

Wright has used this method for the diagnosis of Malta fever (using Bruce's *micrococcus melitensis*) as well as of typhoid, and recommends it strongly on account of its simplicity. No microscope, no thermostat, no apparatus of any kind is needed except the sedimentation tubes. If dead bacilli be used for the test, as Wright has advised, the procedure is brought within the grasp even of the busy practitioner, who has only to keep by him a flask of the harmless dead organisms, which will last indefinitely, and a few sedimentation tubes. The test "performs itself" without labor or watching. (See below, page 30, on the use of dead bacilli.)

To Durham and most of those who have tried Wright's method it has appeared useful chiefly with very powerful sera, such as those of highly immunized animals. With ordinary typhoid blood the sedimentation is often not complete, and the difficulties of securing an aseptic serum are considerable.

III. The Reaction with Dried Blood.

In Widal's original communication he reported that he had found the agglutinating power persisting in the dried blood of a typhoid fever patient even four months after the blood was drawn. He believed that the agglutinative power of the dried blood was, however, somewhat less intense than that of the fluid blood or serum. He did not attempt to apply the method of using the dried blood in diagnosis, and it is to Johnston of Montreal that the credit belongs for having applied the dried blood method to public-health work. For this it is especially convenient, owing to the ease of transmitting dried blood by mail and the comparative difficulty of sending fluid blood.

The blood is to be collected on glass or glazed paper, never on unsized paper, or on anything into which the blood can soak. When thoroughly dry and protected from injury, it can be transported or preserved for an indefinite period without decomposition or loss of its agglutinating power. When we are ready to test it, the blood is simply to be dissolved in water and the mixture of blood and water added to the culture in the way above described. If the drop of blood is dried on a glass slide, the whole process of dissolving and mixing can be carried out on the slide. If paper is used to collect the blood, the dried drop can be cut out with a pair of scissors and rubbed up in a watch glass or in a test tube with one drop of water. When the blood is dissolved ten drops of culture are added, and the examination is carried out in the ordinary way.

More exact dilutions can be made by collecting the blood in the eye of a wire loop of a given size (*e.g.*, 2 mm. inside diameter), and putting on a bit of glass or paper as much blood as will come out of it. When we come to dilute, ten or more drops of water can be similarly transferred to a glass slide with the same wire loop and mixed with one of the blood drops. Under these conditions, the blood and the water being collected in and deposited from the same loop, accurate dilutions can be made.

Block (*Journal of the American Medical Association*, June, 1897) reports the use of the v. Fleischl hæmoglobinometer as a means of diluting dried blood.

A measured amount of blood is dried and then mixed with ten times the amount of culture. The tint of this mixture in the hæmoglobinometer scale is then fixed. Thereafter any specimen of blood to be tested has only to be diluted until the tint is the same as that previously proved to correspond to a 1:10 dilution.

When one comes to the microscopic examination of the mixture so prepared, the pseudo-clumps produced by patches of fibrin with bacilli entangled in them must be carefully differentiated from the true clumps which are composed wholly of bacilli. Biggs and Park state that in non-typhoid blood any slight tendency to clumping which may exist is greatly increased by the presence of fibrin masses in which the bacilli become entangled.

The possibility of puzzling pseudo-reactions with dried blood, and Johnston's method of preventing them, will be discussed when we come to consider the preparation of the test culture.

The method as here outlined has been extensively used only in this country. Pfuhl,[1] Pick,[2] and v. Ordt[3] have applied it in small groups of cases, but in this country the boards of health of every large city except Boston have promptly taken up

---

[1] Pfuhl: Centralbl. für Bacteriologie, 1897, p. 52.
[2] Pick: Wien. klin. Woch., 1897, p. 82.
[3] Ordt: Münch. med. Woch., 1897, p. 327.

Johnston's procedure and applied it successfully in a large number of cases as a measure of public sanitation. Thus Wilson and Westbrook in Minnesota, Biggs and Park in New York, Abbott in Philadelphia, Gehrmann and Wynkoop in Chicago, have investigated the cases of any physician who would send samples of dried blood, and a great body of valuable information has been accumulated. In J. C. Da Costa's article, published in August, 1897, are collected over one thousand nine hundred and fifty cases tested by American observers alone, using the dried-blood method. Since that time many more cases have accumulated.

The happiest results have been obtained by Wilson and Westbrook, who have made fourteen hundred examinations in nearly seven hundred cases of typhoid, using the dried blood only, with the following modifications of Johnston's technique:

With each outfit for collecting the blood there goes: (a) A bit of platinum wire (No. 19 gauge) having one end bent into a loop whose inside diameter is 2 mm. This loop is used to collect the blood, and by its means several drops are deposited on (b) a bit of aluminum foil (No. 40 gauge, 5 cm. square). This foil is rolled up, after the blood has dried upon it, in such a way that the blood cannot fall out or get contaminated. From this bit of foil the blood readily flakes off when manipulated at the laboratory.

The next process is to weigh out, with a delicate balance, 1 mgm. of dried blood and 200 mgm. of distilled water. These ingredients when mixed give us an exact dilution of 1:200 by weight which corresponds to a 1:50 dilution by volume, since blood loses three-fourths of its weight on drying. A hanging drop of the diluted blood is then inoculated with a trace of solid agar culture or bouillon culture and examined; the time limit is two hours.

The results obtained by this method in the enormous number of cases to which it has been applied are apparently more satisfactory than those obtained by any other observer.

On the other hand, we have the authority of Wyatt Johnston (whose experience reaches above one thousand cases) *against* the importance of accurate dilutions for diagnostic purpose. In a recent letter to me (July 20th, 1898) he says: "I can't say that I have been able to make any more positive diagnoses with quantitative methods than with qualitative. I have so far failed to encounter any reaction which I consider typical under conditions that exclude the possibility of a typhoid infection, *although many hundred cases were examined* with this object in view. *In from one to two per cent of apparently genuine cases of typhoid fever I find the reaction either absent, slight, or delayed.*"

Widal prefers on the whole liquid to dried blood. He thinks the former is somewhat more potent and preserves its potency undiminished for longer periods.[1] Yet he found a marked clumping power persisting after six months in a specimen of dried blood. Pfuhl (*Centralblatt für Bacteriology*, 1897, No. 2) has used the dried blood successfully, but thinks that it loses its potency to a certain extent after two or three days. Delepine (*Lancet*, December 5th, 1896) has noticed the same thing. Brannan (New York *Medical Journal*, March 27th, 1897) finds many doubtful or partial reactions—*i.e.*, loose, small clumps without loss of motility—or late reactions, *i.e.*, those appearing after thirty minutes.

My own experience with dried blood in about one hundred and fifty cases has been entirely favorable. I have followed Johnston's methods, and have found no difficulty in getting as accurate results as with liquid blood or serum. (On the accuracy of the serum test in general, see page 63.)

[1] Johnston, on the contrary, thinks dried blood more potent than fluid serum.

CHAPTER III.

THE CLUMP REACTION WITH DEAD BACILLI—MEASUREMENT OF CLUMPING POWER.

ONE of the most astonishing variations in the method of producing the clump reaction was brought to light when Bordet noticed that cholera vibrios were still susceptible to clumping by their own serum[1] after having been killed by chloroform vapor. Widal and Sicard found subsequently that the same is true of typhoid bacilli killed by heat or formol. Thus, bouillon cultures of Eberth's bacilli, after a half-hour's exposure to a temperature of 100° C., are quite dead, and have also ceased to be susceptible to the clumping action of any but the most powerful sera. But if the heat is not run higher than 60° C. during the half-hour, the bacilli (according to Widal) preserve undiminished their susceptibility to agglutination, although they are quite dead and incapable of growth on culture media.

Formol—one drop to one hundred and fifty drops of bouillon culture—likewise "embalms" the bacilli without destroying their sensibility to the clumping power, and the clumps formed among them by the action of homologous serum[1] are in all respects like those formed by living bacilli. Widal kept cultures of typhoid bacilli so embalmed by formol for five months at laboratory temperature, the mouths of the tubes being stoppered with rubber. On these cultures the sera of various typhoid fever patients were found to have a clumping power quantitatively as great as on young live cultures; that is, the sera reacted in as high a degree of dilution as with the live cultures. The dead bacilli settle to the bottom of the tube in the course of time, but can be easily distributed throughout the bouillon again by shaking the tube when we are ready to use it. Of course cultures so embalmed are very safe from contamination.

[1] By "own serum" is meant serum from a case of infection due to the bacillus in question.

V. de Velde[1] has confirmed these results in the main. He used corrosive sublimate, carbolic acid, or thymol as preservatives, and found them as good as formol. Wright and Semple[2] utilized dead bacilli in the diagnosis of typhoid and of Malta fever, and reported that they acted as well as live bacilli. C. Fraenkel (*Deutsche medicinische Wochenschrift*, April 15th, 1897) was equally successful.

J. L. Miller,[3] on the other hand, tried the serum of three typhoid cases on bacilli killed by formol, and got no reaction. M. W. Richardson tells me that he has been equally unsuccessful with the use of dead bacilli. He found them very slow and uncertain in their reactions.

Masius[4] found that ether, corrosive sublimate, chloroform, thymol, carbolic acid, or a temperature of 60° C. could be used to kill the bacilli without impairing their clumping. He succeeded in obtaining clumps visible to the naked eye, using dead bacilli, even with a dilution of 1 : 15,000. Microscopically clumping could be seen in this case in dilutions as high as 1 : 20,000. On the whole he *prefers* and advises the use of dead bacilli rather than live ones, on account of the variability of the latter according to conditions of environment and race. Widal has suggested the same method of avoiding such variation, but apparently has not availed himself of it.

Measurement of Clumping Power.

Widal has made extensive studies regarding the amount of agglutinating force present in different sera, and at various stages in the course of cases of typhoid fever. His method of procedure was originally as follows: "Prepare a series of sterile tubes containing respectively 1 c.c., 2 c.c., 3 c.c., 4 c.c., etc., of bouillon. Add to each a drop of the serum under investigation and a trace of typhoid culture, and leave them several hours in the thermostat at 37° C. If the tube containing 3 c.c., for example, becomes clear, with a precipitate of masses of bacilli at the bottom, while the tube containing 4 c.c. remains as diffusely turbid as ever, the clumping power of that serum is somewhere between 1 : 60 and 1 : 80. In typhoid cases at the height of

---
[1] La Sem. Méd., 1897, p. 114.
[2] British Medical Journal, May 15th, 1897.
[3] Chicago Medical Reporter, May, 1897.
[4] La Sem. Méd., 1897, p. 114.

the disease, the clumping power often reaches 1:100 and often higher."[1]

Stern[2] and Delepine[3] used the quick method in the measurement of clumping power and reported cases of higher potency than any obtained *up to that time* by Widal with his first technique. Widal thereupon adopted their methods and proceeds now in the following way:

"Have on hand a large bulk of bouillon culture, so that the tests made at the different stages of a single case may be strictly comparable one with another. For the same reason all tests should be made with cultures of the same age [and from the same stock culture], incubated at the same temperature. I begin usually with an ordinary 1:10 mixture of serum and young culture and make a microscopic examination of such a mixture between slide and cover. This preliminary examination serves as a guide and enables one after a little practice to tell approximately whether the clumping power is feeble, moderate, or intense.

"Now suppose that the agglutinating power is moderate. We begin by making two dilutions of serum with young culture, one 1:50, the other 1:100 [usually a medicine-dropper is used to measure the proportions], the drops are of course to be of equal size so far as possible.

"If a drop of the 1:50 mixture shows no clumping after from fifteen to thirty minutes, we dilute 1:40, 1:30, or 1:20.

"If, on the other hand, the 1:50 solution clumps readily while the 1:100 mixture shows no tendency to clump, tests are made at 1:60 and 1:80. If there is clumping in the 1:100 mixture we try 1:200, and so on till a dilution is reached which shows no tendency to clumping, and *no agglutinative centres* in a slide-and-cover preparation *two hours old*. . . .

"If the agglutinating power is very high, it is convenient to make dilutions of the serum with sterile bouillon and then add this mixture to the culture. Thus, for example, if a drop of serum be added to one hundred drops of sterile bouillon, and if then a drop of this mixture produces clumping in ten drops of bouillon culture, the agglutinative power is 1:1,000."

As Widal himself admits, it is difficult to fix the limit of

---

[1] Widal: Annales de l'Institut Pasteur, May 25th, 1897.
[2] Stern: Centralb. für innere Medicin, 1896. No. 49.
[3] Delepine: Lancet, December 5th and 12th, 1896.

clumping power exactly. It will be noted that he uses here a *two-hour time limit,* and expects only "agglutination-centres," and not tight clumps at the higher dilutions. [Stern[1] likewise uses the two-hour time limit, but he also puts the preparation in the thermostat at 37° C.] Yet Widal says: "If one goes on diluting the serum beyond the point at which the agglutination centres cease to form, one can still see that the serum exercises a certain influence on the bacilli. They tend to approach each other and lose some of their motility, yet the phenomena are not definite enough to record. Thus, for example, a serum the limit of whose agglutinative power is 1:7,000 exercises *a certain influence* upon the bacilli even at 1:8,000 or 1:10,000, but does not clump them, 'Le dernier terme de cette influence est une sorte de *sideration des microbes.*'"

Clumping is favored by the conditions existing on a slide-and-cover-glass preparation. For example, Widal found that a mixture of highly diluted serum and culture which had been twenty-four hours in the moist chamber at 37° C. showed under the microscope no clumping. But after several minutes' sojourn between slide and cover clumping began to take place. Simple contact of serum and culture in the moist chamber was insufficient to bring about clumping even in twenty-four hours, but the same mixture spread out in a thin layer between slide and cover, and more or less exposed to the air at its edges, showed agglutination within a few minutes. It is obvious, therefore, that any such differences in the mode of making a slide-and-cover specimen as tend to delay evaporation will retard agglutination. Thus, if the layer of fluid between slide and cover is unusually thick, clumping is delayed. If a cover-glass is tilted so as to make the layer of fluid deeper at one point than at another, the bacilli gather at the shallowest point, and clumping is further advanced there. Similar gatherings occur round any bubbles of air that may occur in the preparation. It has been already mentioned that in hanging-drop specimens where the air is excluded agglutination is relatively slow.

These facts teach us that in measuring clumping power "*il ne faut pas se perdre dans l'appréciation des nuances; on ne doit jamais rester sur une hésitation*" (Widal). Indeed, it seems to me that these words apply to all phases of serum diagnosis. Doubtful cases should be classed as negative.

[1] Stern: *Loc. cit.*

By this latter method of Widal's we can dispense with all precaution as regards asepsis, and no special instruments are needed. The blood can be taken from the ear in the ordinary way, since

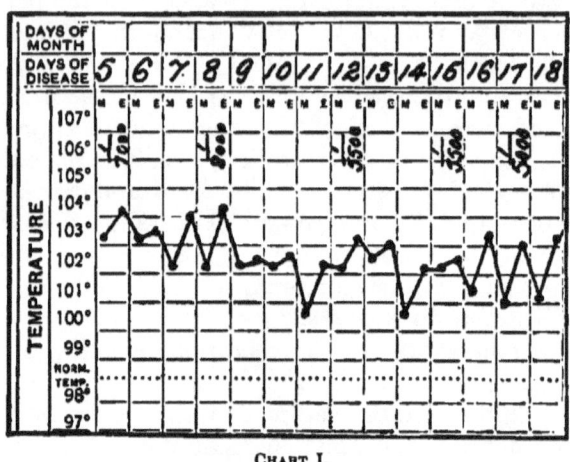

Chart I.

clumping goes on as usual even in contaminated mixtures, and in two hours no considerable contamination occurs. Widal sent

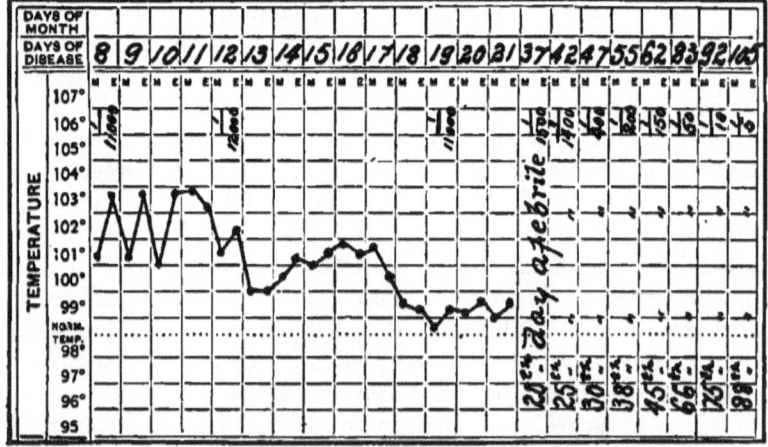

Chart II.

one of his specimens of blood on a ten days' journey after measuring the extent of its clumping power. The tube in which it was transported was *not* sterile, and was sealed with an unster-

THE CLUMP REACTION WITH DEAD BACILLI.

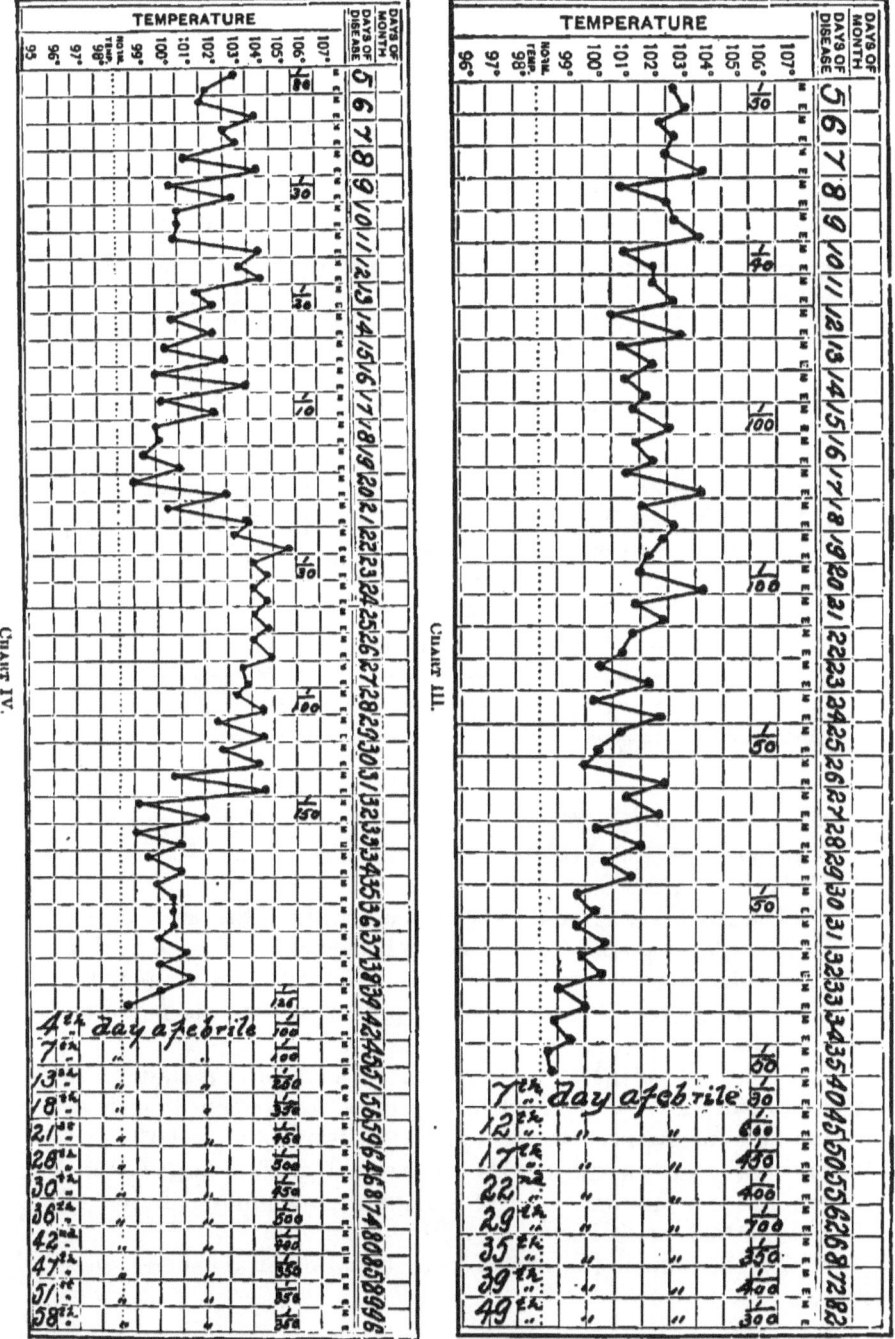

Chart IV.   Chart III.

ilized cork, yet when Widal tested it for the second time after the journey he found the agglutinative power as great as before.

The agglutinative power of typhoid blood varies a great deal in different cases and at different times in the same case, but as yet the causes of these variations are entirely unknown. Widal called a serum "very feeble" if the limit of its power is under 1:100; "feeble," if the limit of its power is between 1:100 and 1:200; "moderate," if the limit of its power is be-

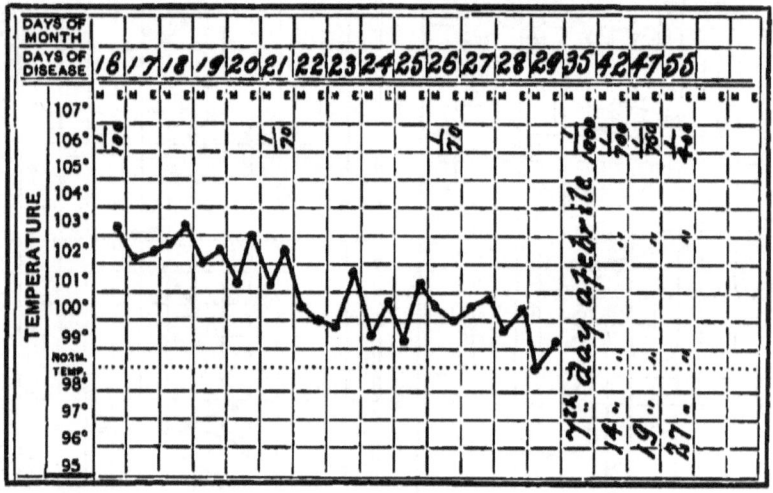

CHART V.

tween 1:200 and 1:500; "intense," if the limit of its power is between 1:500 and 1:2,000; "very intense," if the limit of its power is over 2,000.

The most potent sera do not always occur in the severest nor in the mildest cases. Sometimes it is highest at the beginning of the illness, sometimes at its close. In fatal cases the potency of the serum may fall as death approaches, but this is not always the case.

The most potent serum yet reported, so far as I am aware, is that obtained by v. de Velde after subjecting a horse to a gradual immunizing process for a period of two years. At the end of this time, the serum of this horse would produce large clumps within forty minutes, even at the almost incredible dilution of 1:1,000,000, or .001 mgm. to a litre of culture.

The accompanying charts from Widal illustrate some of the

points made by him. In Case I. we have an intense reacting power in the serum from start to finish, *i.e.*, from the fifth to the eighteenth day; on the latter date the patient died. There is, as will be noticed, a slight decline in the intensity of the reaction as the end draws near; 1:7,000 on the fifth day, 1:5,000 on the day of death. There were no stupor nor signs of severe toxæmia in the earlier days of this case, which appeared to be a mild one at the outset. Yet death occurred from toxæmia without complication.

On the other hand, in Case II. in which the agglutinating power was greater than in any other observed by Widal, the clinical data indicated an infection of moderate severity, the temperature reaching normal during the third week in bed.

There was but mild delirium, lasting a few days only, and convalescence was rapid. This case is notable not only for the remarkable agglutinating power of the serum, but for the rapidity with which this

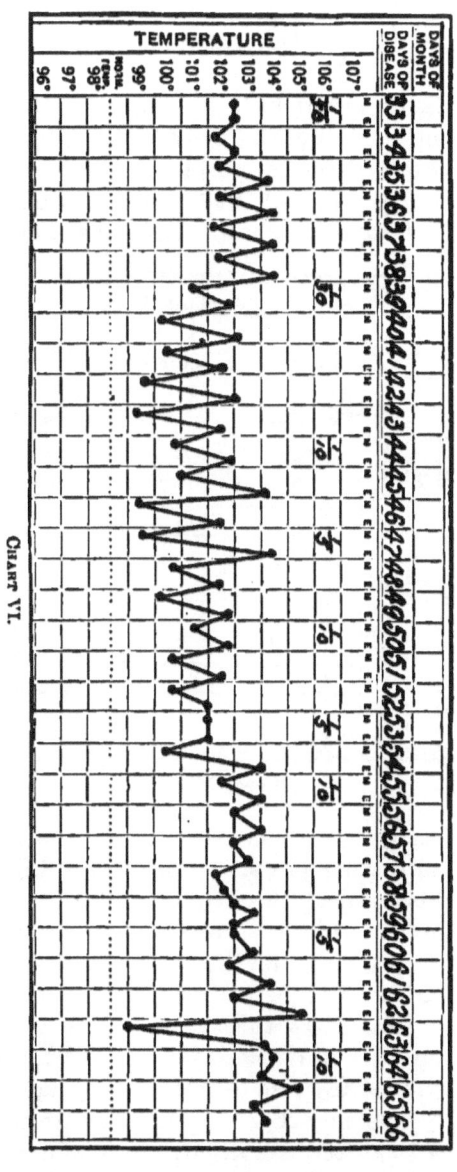

power was lost in convalescence. A little over two months after the temperature reached the normal point, the serum had ceased to show any clumping power when diluted more than 1:10, and two weeks later still it was gone.

Case III. illustrates the precise opposite of the condition just described. During the fever and for some days after the temperature had become normal, the agglutinating power was distinctly less than in most typhoid cases. Yet between the seventh and twelfth day of normal temperature, the agglutinating power of the serum jumped from 1:30 to 1:600, and five weeks later it was still 1:30. It is very hard to account for these facts.

Case IV. shows how the agglutinating power may be stronger in relapse than in the original attack, and stronger still in convalescence.

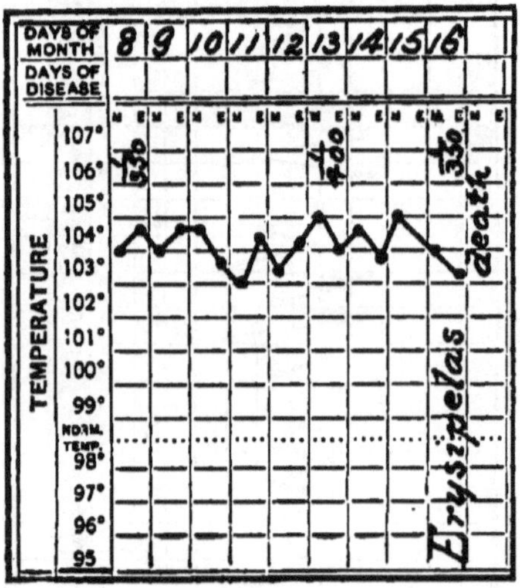

Here the serum grew steadily less potent as the original attack progressed, and one might suppose that there was some causal connection between this fact and the occurrence of relapse. But in Case V. we see a similar fall of clumping power parallel to the fall of temperature, yet no relapse followed.

CHART VII.

Case VI., which ended fatally, is a demonstration of the fact that the degree of agglutinating power is not always high in severe cases. It is in this respect the opposite of Case I. Moreover, in Case IV. the reaction was repeatedly absent at the standard dilution (1:10), and after the first week was very feeble at all times.

Case VII., a fatal case, complicated by erysipelas, shows that in fatal cases the serum reaction may be neither high (as in Case I.) nor low (as in Case VI.), but simply moderate, and without any rise or fall as death approaches.

The only conclusions to be drawn from these charts, so far as I can see, are negative. There is no apparent rule or reason in them. They fit no theory of the nature of the process, and give us no hints for prognosis. The question of the possible prognostic significance of the serum will be discussed later (see page 113).

Yet although Widal's researches in the measurement of agglutinating power have given us no general rules as to the agglutination curves, they have taught us, when taken in connection with Stern's similar investigation, that the potency of typhoid serum is much greater than had been supposed. They show us that *in the great majority of cases the agglutinating power of the serum is over 1:50, that 1:500 is by no means rare, and 1:12,000 possible. Occasionally the serum never becomes potent in dilutions over 1:10; but this is rare.*

CHAPTER IV.

THE PREPARATION OF CULTURES FOR USE IN SERUM DIAGNOSIS. THE CLUMP REACTION WITH OTHER BODY-FLUIDS, AND WITH FILTERED CULTURES. AUTO-SERUM REACTION. REACTIONS WITH CHEMICAL REAGENTS.

It goes without saying that the cultures used should have been identified as positively as possible. In the case of typhoid cultures, with which we shall deal chiefly in this section, all the recognized tests for the identification of Eberth's bacilli should be applied. So far all writers are agreed.

When we come to go beyond this point, very stubborn differences of opinion appear. A great deal of discussion has dealt with the question: "Do different races of Eberth's bacilli (equally well identified) differ in their sensitiveness to the clumping action of typhoid serum?"

There is no question whatever that the degree of this sensitiveness is greatly affected by the conditions of nutriment, temperature, etc., under which the organism is grown. The only problem is: Are *all the differences* in the sensitiveness of different cultures to serum explainable by an appreciation of the conditions under which they have been cultivated, or do different races possess certain inborn differences in sensibility which cannot be produced or eradicated by such conditions as nutrition, temperature, degree of alkalinity or acidity of culture media, etc.?

1. *Differences Between Different Races.*

Durham[1] tested nineteen different cultures brought from various parts of Europe, and found some differences in the rapidity with which a given serum would produce clumping in different cultures.

Widal, working with twenty-six different races of bacilli,

---

[1] Durham: Journal of Path. and Bact., July, 1896.

some of them from America, could not satisfy himself that there were any real differences between them. Sometimes with a given serum he would get quicker reactions on one culture than on the others. But serum from another case would act differently. Cultures very slow in reacting to serum A may be unusually quick in their response to serum B. In fact, Widal says that he has no preferences among the twenty-six cultures which he has in his laboratory, and uses them indifferently. He has among them some fresh from the organs of a patient recently dead from typhoid fever, others which have been transplanted daily for several months, and one which was recently transplanted after having been shut up away from light and air for five years and a half. But he finds no considerable differences in their susceptibility to typhoid serum. In three cases in which he compared the effect of patients' serum on the bacilli isolated from their own stools he found that the serum produced no greater clumping with these particular races than with stock laboratory cultures. One culture would perhaps clump at 1:30, and no higher with a given serum, while another culture might agglutinate at 1:40. . In no case did he find a greater variation than this.

C. Fraenkel's conclusions are similar.

Achard and Bensaude[1] tested twenty different races of typhoid bacilli from various sources—all carefully identified. In the great majority of cases they found no considerable difference between the different cultures, provided they were grown and tested under similar conditions. In one case, however, they found that on the eighteenth day of normal temperature the serum would clump one culture out of nine tried, but remained without effect upon the others.

V. de Velde[2] examined twenty cultures, among which he found one which scarcely reacted at all to typhoid serum (two cases). He used altogether the macroscopic method, and his results are thereby invalidated to a considerable extent (in my opinion).

The following table from V. de Velde illustrates the differences found by him:

[1] Achard and Bensaude: Presse Méd., November 25th, 1896.
[2] V. de Velde: Bull. de l'Acad. roy. de Belgique, March 27th, 1897.

TABLE II.—SERUM OF PATIENT "X," DILUTED TO 1:10.

Name of Culture and Tube Mark.		Half-Hour After Mixing.	One Hour After.	Two Hours After.	Twenty-four Hours After.
"Lille"	A.	?	+	+	+
	B.	+	+	+	+
"Berlin"	A.	0	?	?	0
	B.	0	0	0	0
"Paris"	A.	0	?	+	+
	B.	0	?	+	+
"10"	A.	?	+	+	+
	B.	+	+	+	+
"14"	A.	+	+	+	+
	B.	+	+	+	+

Masius[1] tested a large number of cultures and found that some of them reacted much better than others to a given serum. He believes that this fact has led to many untoward results, and advises the use of such cultures only as have been proved to be very susceptible to typhoid serum.

Mills used twenty-eight different races and found considerable differences in their susceptibility to serum. He considers that the rate of clumping varies inversely with the virulence.

Biggs and Park used throughout the experiments reported in their first communication (*American Journal of the Medical Sciences*, February, 1897) a culture received from Pfeiffer. This culture they state to have been far more virulent and more easily clumped than any obtained from other sources.

Förster (*Zeitschrift für Hygiene*, etc., Vol. XXIV.) used nine different races and detected no considerable differences in their susceptibility to serum.

M. W. Richardson has worked with fifty-five different races of typhoid bacilli, most of them isolated by him from the urine, stools, or tissues of patients at the Massachusetts General Hospital. He has tested all of these cultures with a single serum of known strength, which he has preserved by soaking it into paper and drying it there. Fifty-one out of fifty-five of these races have shown practically the same degree of sensitiveness to typhoid serum, *i.e.*, clumping has been complete within five minutes with a mixture corresponding to a 1:10 dilution.

The other four cultures (two of them from urine, one from a gall-bladder many weeks after the fever, and one from a typhoid

[1] La Sem. médicale, 1897, p. 122.

stool) were much slower in reacting. This he thinks may have been due to the conditions under which they had grown up, since passing them through a guinea-pig restored their susceptibility in two cases to a considerable extent. All of the four refractory cultures were practically non-motile, motility and susceptibility to clumping running nearly parallel in Richardson's experience (personal communication).

I have myself used some twenty or thirty different races, and have not been able to determine any considerable differences in their susceptibility to serum except such differences as were explicable by difference of conditions.

On the whole, therefore, it appears that the majority of observers find no considerable difference in the availability of different races of Eberth's bacilli for the purposes of serum diagnosis, and on preparing our cultures it makes no considerable difference what race we start with. *The crucial point is the environment* in which our test culture is grown.

2. *Effect of Environment upon Cultures as Regards their Availability for Serum Diagnosis.*

All writers are agreed that a culture may be fit or unfit for use in diagnostic work according to the condition of its growth. Yet when we come to go farther and inquire, "*What conditions produce the most suitable cultures for diagnostic work?*" we find the most confusing diversity among the reports of different observers upon this point. Half the observers recommend the "most virulent" and actively motile cultures as the best, while the other half prefer "attenuated" cultures. The source of these differences is not difficult to find. Those who recommend the fresh, "virulent," swiftly darting bacilli as the best, mean by "the best" those which are most sensitive to agglutinative serum; while those who use attenuated cultures do not wish to get the most sensitive bacilli, nor those which clump the easiest. To the latter observers it appears that a culture may easily be worked up into a *supersensitive* or *hyperæsthetic* state in which clumping occurs *too readily*. "Too easily" means that even normal serum may clump such a culture so that with it there is nothing distinctive of typhoid in the fact of clumping.

The great majority of writers hold that the more virulent and motile the culture, the more easily it is clumped. But Mills (Moscow Congress, August 23d, 1897), Kolle (*Deutsche medicinische Wochenschrift*, 1896, p.

152), Weaver (Chicago *Medical Recorder*, May, 1897), Stern (*Centralblatt für innere Medicin*, December 5th, 1896), and others believe that the more attenuated the culture the more susceptible to serum, the very opposite of the findings of Johnston, McTaggert, and the great body of those who have studied the subject. The explanation of these differences lies, I think, in the lack of any standard tests for "virulence" and "attenuation," except the actual use of animal experimentation, which few observers have resorted to. As a rule "virulence" and "attenuation" are very loosely used by writers on serum diagnosis.

Those who deny that a culture can be too sensitive begin by freshening up the culture with which they intend to work by repeated transplantations. There is no question that the motility of many races can be increased by daily transplantations. Thus, for example, I remember a culture obtained from the spleen of the dead fœtus of a woman who had miscarried during an attack of typhoid fever. These organisms fulfilled all the standard tests for the identification of Eberth's bacilli except that their motility was very slight. After a few transplantations in the thermostat on agar-agar, these bacilli grew nearly as motile as those ordinarily seen. Their response to the action of a given typhoid serum increased parallel to their increase of motility. Probably the conditions of nutrition in the spleen of the fœtus were sufficiently unfavorable to deprive the bacilli for the time of their naturally active locomotive power, which was easily restored by a proper diet. This was a case in which only one colony developed in the culture tubes, although a large amount of material from the spleen had been spread upon them.

I have seen similar behavior in a race of bacilli cultivated from a single colony, which was the only result of a culture from the spleen of a typhoid patient who died of perforation in the fifth week of the disease. Most of the ulcers were wholly healed and the spleen was not enlarged.

Fresh cultures from the spleen of typhoid patients, living or dead, are usually very motile and sensitive to serum. But if kept in the thermostat, and not transplanted, they grow into long, sluggish involution-forms very hard to clump. Daily transplantation in bouillon prevents this partially, yet gradually diminishes the motility of some races provided the cultures are kept at 37° C. But if the cultures are kept at room temperature they grow far more slowly and need transplanting only once in two or three days. Under these conditions they maintain their motility and their sensitiveness to typhoid serum for very long periods. If thermostat bouillon cultures are used, they are most available when from ten to fourteen hours old, that is, before the turbidity of the bouillon has grown marked, and before any sediment or scum has accumulated. Twelve-hour cultures at 37° C. correspond roughly to thirty-hour cultures at room temperature. In general I think it will be found that the more motile the culture the more easily it is clumped.

Most of those writers who have worked with fluid bood or serum have preferred to have the culture as actively motile and easily clumped as possible, avoiding the pseudo-reactions which occasionally occur with non-typhoid blood by dilution of the serum.

Johnston, on the other hand, found that with the dried-blood method some specimens of normal blood would clump actively motile cultures even in a dilution of 1:40. He seeks to avoid such misleading pseudo-reactions, not by dilution of the serum, but by attention to the condition of the culture. He reduces the motility of the bacilli by keeping his stock agar-cultures at room temperature, and transplanting them only once a month. This "attenuates" the bacilli so that in twenty-four-hour bouillon cultures from such an agar culture the bacilli show a slow, gliding motion, but no darting motion, no furious activity.

He also pays great attention to the *reaction of the bouillon* into which his transplantations are made. In general it is laid down that bouillon should be neutral. Johnston finds that bouillon just on the verge of litmus acidity works best. Such bouillon gives no blue to the red paper; 3.5 per cent. of normal alkali is needed to make this bouillon neutral to phenolphthalein. Such bouillon enables him to dispense altogether with precautions as to dilution and time limit.

Block (*British Medical Journal*, December 18th, 1897) makes the following observations for the preparation of the bouillon for cultures:

I. It has been shown that hydrochloric acid used in the neutralization of bouillon exerts no influence upon the clumping process. But if the reaction is carried beyond the point of neutralization, and caustic soda is added for correction of the acidity, we have a source of error in case undistilled hard water is used in any part of the test. Calcic carbonate may then be precipitated, a substance which, according to Malvoz (*vide infra*, page 54) has a marked agglutinating power.

II. In alkaline media spontaneous clumping is especially apt to occur.

III. If too frequently transplanted, cultures are obtained at last in which spontaneous clumping, scum, and sediment occur.

The above conditions are therefore to be avoided.

Biggs and Park (*American Journal of the Medical Sciences*, February, 1897) tried Johnston's method of making his bouillon transplantations from old stock cultures kept at room temperature. But in their hands cultures so prepared were not reliable, and gave rise to many pseudo-reactions. They found the most virulent and active cultures the best.

Stern (*Centralblatt für innere Medicin*, December 5th, 1896) makes use of two cultures in every case—one as fresh as possible from an autopsy, and one that has undergone many transplantations.

Dr. Leary, of the Boston City Hospital, also keeps his cultures at room temperature but transplants more frequently than Johnston does. Indeed, it is sometimes impossible to leave an agar culture for a month at room temperature without killing it. Dr. Leary believes that bacilli grown at room temperature and transplanted every few days are more normal than those grown under the hot-house conditions of the thermostat. The climate of the thermostat he thinks produces after a time an enervated, overstimulated race whose reactions are less to be depended on than those of a race brought up in the more normal environment of room temperature. He thinks that thermostat cultures react more quickly than those grown at room temperature—too quickly in fact, since they may give rise to pseudo-reactions with non-typhoid serum (time limit one hour). Cultures grown at room temperature show scantier and larger bacilli, and do not need such frequent transplantations.

My own practice has been to transplant wholly from bouillon to bouillon, and to keep the cultures wholly at room temperature. Such cultures need transplanting about once in three days, though more frequent transplantations are of no harm.

The use of suspensions or emulsions of bits of solid culture in bouillon was at one time preferred by Durham and others to that of bouillon cultures. By most others it has usually been valued only as a convenient makeshift in case we have no bouillon cultures at hand and cannot wait for them to grow. I have already referred to the difficulty and the importance of thoroughly dissociating the bacilli in the suspensions. Nevertheless, Wilson and Westbrook, and C. L. Greene have obtained excellent results with the use of suspensions, and there is no doubt of their usefulness in skilled hands.

One point on which all observers are agreed is that the bouillon culture finally used should be *young*, however the stock culture is treated. In old bouillon cultures spontaneous clumping often occurs. This occasionally happens even in young cultures if the bouillon is not right or some other condition is unfavorable; but in old bouillon cultures, *i.e.*, over thirty-six hours in the thermostat, or over three days at room temperature, it is relatively frequent.

Johnston's methods have been followed by many other observers, usually with entire success. Elsberg, Appel and Thornbury, Da Costa, Abbott, Wilson and Westbrook, Gehrmann and Wynkoop, Biggs and Park, and others have used the dried-blood technique in connection with public health work in an enormous number of cases with the happiest results (see page 28).

Summing up the discussion of the preparation of cultures it appears that all observers agree on the following points:

1. Cultures should be free from sediment or pellicle, and only slightly turbid (however this result is attained).

2. The presence of long involution forms, of any tendency to spontaneous clumping, or of non-motile or very sluggish bacilli should lead us to discard any culture for immediate use in serum diagnosis.

3. As a rule such cultures can be made available by a few transplantations and a short sojourn in the thermostat.

The only points of difference are as to whether the bacilli are to be as motile as possible, or only moderately motile. Those who prefer the former escape the danger of pseudo-reactions by high dilution of the serum. Those who prefer the latter believe that they can dispense with the dilutions altogether by attending scrupulously to the condition of the culture.

To get a very active culture it is best to use the thermostat, to transplant the stock agar cultures every forty-eight hours, and to use for our test culture a bouillon transplantation about twelve hours old. Cultures recently isolated from the tissues of typhoid patients are usually more active than those which have been subjected for a considerable time to artificial conditions. An increase of virulence, motility, and sensitiveness to serum can usually be produced by passing the bacilli through an animal.

If, on the other hand, we prefer to avoid pseudo-reactions by using attenuated cultures, as Johnston has advised, it is better to keep the cultures wholly at room temperature and to transplant them less frequently.

Preparation of Cholera Cultures.

So far we have spoken wholly of the preparation of typhoid cultures. As regards the cholera vibrio, Bordet has noted considerable variation in the behavior of the different races under the influence of a given serum, and Achard and Bensaude have called attention to the difficulties caused by the pellicle. To these points we shall return in the section on Cholera.

The differences between different races of pneumococci, of streptococci, of colon bacilli, and of the proteus group are very considerable, as we should expect. They will be discussed under the sections relating to these organisms.

THE CLUMP REACTION WITH OTHER BODY FLUIDS.

Besides the whole blood (fluid or solid), the serum and blister fluid, various other juices and secretions can be used in a similar way to demonstrate the presence of the poison of typhoid fever in an individual's tissues.

The blood is, to be sure, more strongly agglutinative than any other body fluid, and its plasma is even more potent than its serum, yet it has no monopoly of the clumping power, since, as Widal has shown, "the various membranes of the organism permit more or less diffusion of the agglutinating material contained in the blood."

1. *In pus* Catrin found the reaction present in man (two cases) and Achard and Bensaude observed the same fact in animals. Widal succeeded in immunizing an ass so strongly with Eberth's bacilli that its blood would clump typhoid bacilli when diluted $1:14,000$. In some pus obtained from this same animal he found an equal potency. A flask of this pus was preserved by him for fifteen months, at the end of which time the supernatant liquid clumped at $1:13,000$. This pus was full

of live typhoid bacilli, their presence seeming in this case to have no considerable effect.

2. *In the urine* the reaction is very inconstant, as Widal showed in 1896. "It appears and disappears from day to day, almost from hour to hour, without any obvious reason for the variations." In a goat so strongly immunized that its blood possessed a clumping power of $1:8,000$ Widal and Nobécourt found that the urine had no clumping power. In a typhoid-fever patient whose blood had a potency of $1:800$ the urine occasionally would react at $1:10$. No reactions were obtained in any case except in feeble dilution. On the other hand, the urine of non-typhoid subjects has occasionally been found active.

Bormans (*La Riforma Medica*, 1896, Nos. 274 and 275) has found the reaction in the urine of several typhoid patients, and failed to find it in his control cases. He warns us of the need of using only catheter specimens of urine in women.

3. *In the stools* of typhoid patients with diarrhœa, Achard and Bensaude found clumping power only when blood and pus were present. Solid stools brought away by enema were without agglutinative power.

The same observers tested the stools of six cholera patients and found that they were without effect upon cholera vibrios even in dilutions as feeble as two or three parts to ten. Block found the reaction positive with the stools in two cases of typhoid.

4. *In the milk* of women attacked with typhoid during lactation, Achard and Bensaude were the first to discover that the reaction is often quite marked, while the milk of many healthy wet-nurses was found to be wholly without clumping power, and the milk of women suffering from other disease was equally negative.

They noted that only fresh and pure milk from the typhoid patients could be used for this test, since notable variations in the clumping power begin to show themselves as soon as the milk has been kept a short time. These varieties are not due simply to changes in reaction, since the artificial production of moderate acidity or alkalinity does not affect the clumping power of milk to any extent. On the other hand, milk passed through a porcelain filter loses its clumping power, and a temperature of $120°$ C. maintained for twenty minutes likewise destroys it. In making the test the milk is simply to be mixed with the culture $1:10$, as in the case of blood.

Achard and Bensaude got the same results with the milk of rabbits inoculated with typhoid bacilli, and a similar effect on cholera vibrios was produced by the milk of animals inoculated with this organism.

Widal measured accurately the clumping power of the serum and milk of a goat artificially immunized with Eberth's bacilli; the fluids were both collected at the same time and on the same day. The serum clumped 1:6,000, the milk only 1:400.

Mossé found a marked agglutinating power in human *colostrum*, confirming Boclère's observation. The reaction was, however, much slower and less marked than in the serum.

5. *Pericardial, pleural, and peritoneal fluid (normal) and œdema fluid* likewise possess the clumping power, though to a considerably less extent than the serum.

In a case tested by Widal the serum was potent at 1:350; the pericardial fluid only at 1:60.

6. *Tears* have been tested by Widal in fourteen cases, in six of which the blood gave marked reaction. Ten of these cases were at the height of the disease, three convalescent, and one had had typhoid seven years before. In four of these cases the reaction could not be produced by the tears even at a dilution of 3 or 4:10. In one case it appeared only at a dilution of 3:10, in three cases at 1:5, and in six at 1:10. With higher dilutions no reaction could be obtained, and in three cases tested post mortem the results were negative. It is an interesting fact in this connection that tears artificially stimulated by ammonia fumes or other irritants appear to be nearly devoid of agglutinating power. Widal found only the very slightest reaction in three out of fourteen cases in which tears were produced in this manner, and then only in dilutions lower than 1:10. It is only with the natural lacrymal secretion such as can be sucked out with a pipette from the internal corner of the eye that the reaction can be obtained.

In six immunized animals the natural tears gave fairly marked reactions. In five healthy persons and three normal rabbits, there was no reaction in the tears.

7. Somewhat similar phenomena are suggested by the case of Ménetrier in which no reaction could be obtained with the serum of a *pleural effusion* complicating a case of typhoid fever, although the normal pleural fluid usually does produce the reaction in typhoid cases examined post mortem. Courmont

found the reaction well marked in serous effusions of all kinds, whether acute inflammatory or other. The inflammatory effusion in Ménetrier's case contained a pure culture of Eberth's bacilli, and it has been argued by Courmont and others that the presence of these bacilli explained the absence of agglutinative power in the fluid. That this explanation is not wholly satisfactory is shown by Widal's case, already quoted, in which the reaction persisted undiminished for fifteen months in a bottle of pus swarming with live typhoid bacilli. In de Rochemont's case of pleuritic effusion complicating typhoid, there was no clumping power in the effusion, though it contained no typhoid bacilli, and the blood was markedly agglutinative. Yet Courmont found that typhoid serum soon lost its clumping power when Eberth's bacilli were grown in it outside the body. Antony and Ferré (*Journal de Médecine de Bordeaux*, 1897, No. 30) reported two cases in which Eberth's bacilli were cultivated from the blood. In neither of these did any agglutinative reaction appear in the serum.

Blumenthal (quoted in *La Semaine médicale*, 1897, p. 152) had a similar experience in one case. These cases will be referred to in more detail later.

8. *Bile* was negative in one case and positive in a second case tested by Widal.

Achard and Bensaude found the bile likewise negative in two cases post mortem. In one of these cases a pure culture of typhoid bacilli was obtained from the bile. In an immunized rabbit they obtained the reaction once. Courmont tested one case in man with positive results (1:10).

Seminal fluid was without power in two cases, and *cerebrospinal fluid* in four cases [1] examined by Widal.

Saliva, gastric juice, and *bronchial secretion* were found to be negative by Achard and Bensaude in several cases.

Natural *sweat* was tested by Thiercelon and Lenoble (*La Semaine médicale*, 1896, p. 496) with negative results in the case of a patient whose blood was strongly agglutinative.

Courmont has made a detailed study of the juices of various organs at autopsy in seven cases, one of which gave the following results:

---

[1] Salrazés and Rivière have recently reported an experiment upon a dog inoculated with tetanus. The cerebro-spinal fluid agglutinated the tetanus bacillus (Soc. de Biol., June 26th, 1897).

Maximum clumping power of
1. Pleural fluid....................1 : 200
2. Heart's blood....................1 : 100
3. Blood of renal vein..............1 : 100
4. Juice of ovary...................1 : 100
5. Peritoneal fluid.................1 : 100
6. Blood of portal vein.............1 : 50
7. Blood of hepatic vein............1 : 10
8. Blood of splenic vein............1 : 10
9. Splenic juice....................1 : 10
10. Bile............................1 : 10
11. Juice of mesenteric glands......1 : 10
12. Pericardial fluid...............1 : 10

The most intense reaction of all was in the peripheral blood. From these facts Courmont concludes (*La Semaine médicale*, 1897, p. 105) that the agglutinating substance is made in the blood, excreted by the kidneys, and more or less destroyed in the liver and spleen and mesenteric glands, *i.e.*, where live bacilli are present.

Double Infections.

Courmont and Martin (*Lyon médical*, 1897, No. 10) report a case of double infection by the diplococcus lanceolatus and the typhoid bacillus. At autopsy the blood and body fluid, together with the juices of the various organs—all showed markedly a power to agglutinate the typhoid bacilli, *excepting the splenic juice*, which showed no clumping power. Eberth's bacilli were obtained in cultures from the spleen. The fluid from a pneumococcus pleurisy which complicated this case agglutinated Eberth's bacilli promptly, though pneumococci were cultivated from the fluid and no typhoid bacilli were found in it.

Achard and Bensaude (*La Semaine médicale*, 1896, p. 393) found that the fluid of a *hydatid cyst* in a rabbit which had been inoculated with Eberth's bacilli had become endowed with marked agglutinative power.

Transmission to the Fœtus or to the Suckling.

The agglutinating power occasionally passes through the placenta to the fœtal blood in the course of pregnancy. Mossé and Daunic[1] found a well-marked clumping power in the blood of a new-born child whose mother had had typhoid fever during the sixth month of pregnancy. The mother's blood, however,

[1] Mossé and Daunic: Soc. Méd. des Hôpitaux, March 5th, 1897.

had a considerably greater agglutinating power than that of the child. Etienne,[1] on the other hand, reports a case of abortion during typhoid fever in which the blood of the mother gave marked reaction, but that of the fœtus none.

Shaw[2] found a marked reaction in the child of a woman who had had typhoid during her pregnancy. The child was six weeks old.

Apert and Charrier (*La Médecine moderne*, November 11th, 1896) searched all the body fluids of a three-months' fœtus born of a typhoid mother whose blood had marked agglutinative power. The fluids taken from the fœtus showed no clumping power, although the placental blood was markedly agglutinative.

Griffith found the reaction still positive in a child two months old, whose mother was in the third week of typhoid at the time of the child's birth.

Chambrelent and St. Phillipe (*Journal de Médecine de Bordeaux*, March 15th, 1897) had a similar experience.

Grünbaum (*Science Progress*, October, 1897) mentions two cases of children born while the mother was sick with typhoid. One child's blood was strongly agglutinative, the other's not.

Stengel (New York *Medical Journal*, 1898, No. 10) describse the case of a woman who had typhoid when nine months pregnant. Her blood showed a positive agglutinative reaction. After two weeks of fever the child was born and showed at birth a temperature of 102° F. This gradually fell in the course of three days, and both mother and child recovered. There was no clump reaction obtainable with the baby's blood.

Landouzy and Griffon (Société de Biologie, November 6th, 1897) report the transmission of the clumping power from a mother to her three-months' infant through the medium of her milk, the mother having been stricken with typhoid three months after parturition.

In Bracken's case (Philadelphia *Medical Journal*, January 8th, 1898) the baby had no fever, but continued to exhibit the reaction when two months old. Here it may have been transferred by the milk, since the child had nursed one day before the reaction of its blood was tested.

Taking all these facts together, it is evident that no rule can

---

[1] Etienne: Presse Médicale, 1896, p. 465.
[2] Shaw: Lancet, August 28th, 1897.

be made as to the passage of the agglutinating power through the placenta or by the milk.

THE CLUMP REACTION WITH FILTERED CULTURES.

In April, 1897, M. R. Kraus[1] reported before the Gesellschaft der Aerzte at Vienna the results of his researches with cultures of typhoid bacilli, or of cholera vibrios from which the bacilli themselves had been removed by filtration. He found that in such germ-free culture media a *specific precipitate* was formed when homologous serum was added, and the mixture left in the thermostat at 37° C. for twenty-four hours. That is, if typhoid serum be added to such a filtered typhoid culture, a flocculent precipitate occurs, while no serum from healthy persons nor from diseases other than typhoid causes any precipitate. Similar specific precipitates were produced by cholera serum in filtered cultures of cholera vibrios, and by plague serum with filtered cultures of plague bacilli.

The precipitate, when examined microscopically and chemically, seemed to consist of fragments of protoplasm from the bodies of the bacilli, these fragments being so minute as to pass through the filter which held back all the intact bacteria. Germ-free toxins, such as that of diphtheria in which there are no fragments of bacilli, give no specific precipitate when diphtheria antitoxin is added to them. Bensaude has confirmed Kraus' results, using filtered cultures of the bacillus proteus.

Widal and Sicard (*La Semaine médicale*, 1898, No. 19) confirmed these results and instituted researches as to the comparative clumping power of a given serum on filtered and unfiltered cultures. Nicolle had previously found that two typhoid sera which had a clumping power of $1:8,000$ and $1:300$ respectively were negative with filtered cultures if diluted more than $1:10$.

Widal and Sicard tried fifteen specimens of typhoid serum of a clumping strength varying from $1:50$ to $1:20,000$ when mixed with young unfiltered typhoid cultures. None of these sera had any agglutinating power with filtered cultures if diluted more than $1:10$. Furthermore, two sera with a normal agglutinating power of $1:400$, and one of $1:200$ would not react at all on filtered cultures even in a $1:10$ dilution. The serum

[1] Kraus: Wien. klin. Woch., 1897, No. 32.

of a long-immunized ass which clumped unfiltered cultures at 1:50,000 was potent only at 1:10 with a filtered culture.

Many specimens of non-typhoid serum were tested on filtered cultures of 1:10 with negative results; but the serum of two cases of pneumonia, and normal serum from a horse and from a rabbit, produced a precipitate in the filtered culture when mixed with it in equal parts.

Widal was never able to produce the precipitate with serum diluted more than 1:30, and only once at that dilution; 1:10 is usually the limit. Two sera with potencies of 1:1,400 and 1:1,500 showed a reaction with the filtered serum as high as 1:20.

"AUTO-SERUM REACTION."

Under this title Mills refers to experiments comparing the agglutinating power of the blood upon the bacillus isolated from that case with its power over stock laboratory cultures. Eight cases were tested. Six of them reacted better on the laboratory cultures than upon the bacilli isolated from their own stools, and in only two were these conditions reversed, a very surprising result, considering Widal's experience with colon bacilli under similar conditions.

Widal himself tested in three cases the comparative effects of the patient's sera on their own bacilli, and on laboratory cultures, and found no special differences.

THE CLUMP REACTION WITH CHEMICAL REAGENTS.

Blackstein discovered that chysoidin possesses the power of clumping cholera vibrios to a marked degree, while, according to him, it has no effect on allied vibrios nor on other bacilli.

Malvoz[1] experimented on typhoid bacilli with a variety of substances: Formalin, hydrogen peroxide, corrosive sublimate, alcohol, safranin, vesuvin, indulin, and nigrosin will agglutinate Eberth's bacilli when mixed with them in equal parts. Mineral acids, carbolic and lactic acids, and chloroform have no effect. Salicylic acid forms very small clumps. Permanganate of potassium makes fair-sized but very loose heaps.

Malvoz has attempted to use this agglutinative action of chemicals for the differentiation of colon and typhoid bacilli, and appears to find a considerable difference in the behavior of these two types when subjected to the action of various chemicals.

[1] Malvoz: Annales de l'Institut Pasteur, July 25th, 1897.

CHAPTER V.

THE ORIGIN AND NATURE OF THE AGGLUTINATING SUBSTANCE.

IT has been already shown that the clumping material or power, whatever its nature, is found in a great variety of secretions, excretions, and body juices, but more especially in the peripheral blood.

The fact that the blood has usually a more intense clumping power than any other tissue or fluid in the body might lead us to suspect that the agglutinative material was evolved first in the circulating blood, and transferred therefrom to the other tissues through which the blood circulates, and to the other juices and fluids which communicate more or less freely with the intravascular blood by osmosis or otherwise. In the absence of conflicting evidence, we may use this as a working hypothesis.

If, then, the clumping power originates in the blood under the influence of the " materies morbi" of the disease in question, we naturally ask ourselves whether it is evolved by the cellular elements of the blood or whether it arises out of plasma itself.

1. To determine this question, Achard and Bensaude mixed blood from a typhoid patient with extract of blood leech to prevent coagulation, and then separated the leucocytes by filtration. The leucocyte mass was then washed with sterile artificial serum until the filtrate gave no reaction with typhoid bacilli. Having thus removed the plasma from the leucocytes, they subjected them to pressure and tested the juice thus squeezed out of them. This juice was found to be without agglutinating power. That this was not due to the leucocytes being dead was proved by the fact that at the end of the experiment the leucocytes were still able to ingest particles of carmine mixed with them.

Widal and Sicard collected typhoid blood in a sterilized collodion tube (coagulation being prevented by the addition of potassic oxalate), drew off the supernatant serum, and then ligatured the tube at the junction of the layer of red corpuscles underneath with the leucocytes above them. The plasma left in contact with these leucocytes for forty-eight hours showed no more agglutinative power than the decanted serum.

Obviously these conditions are highly artificial, and we cannot conclude that because the leucocytes do not produce any clumping material outside the body and under these special circumstances, therefore they do not produce any in their normal environment. This question must be left undecided.

2. As to the *chemical composition* of the clumping substance, we are very much in the dark. We have seen above that with fluids poor in albuminoid substances, such fluids as the saliva or the cerebro-spinal juice, the clumping reaction does not occur. This suggests that the agglutinative substance may be in some way connected with the albuminoids of the serum. This idea receives some confirmation from the results of filtration experiments. It has been shown by Widal and by Achard and Bensaude that most agglutinative fluids (serum, milk, urine) lose their power wholly or in part after passing through a porcelain filter.[1] Now it is well known that the albuminoids are especially apt to be separated from any fluid during its filtration through porcelain. This suggests to us that the agglutinating material may itself be either an albuminoid or a substance prone to cling to the albuminoids during filtration. The same thing is suggested by the results of precipitating the globulins from the serum by the addition of magnesium sulphate. The filtrate left after precipitating the globulins is devoid of clumping power, while the globulins dissolved in water clump very strongly.

In like manner experiments with milk from animals inoculated with Eberth's bacilli showed that the separation of the casein deprived the milk of all agglutinating power, while the casein itself was markedly agglutinative.

The fact that plasma has a greater clumping power than has serum suggests that the substances concerned in the formation of fibrin possess a part of the clumping material.

Widal finds that in a blood whose serum has a clumping power of 1:40, the plasma is often potent at 1:60.

Precipitating the fibrinogen from plasma and redissolving it in water gives us a markedly agglutinating fluid. But the filtrate is also potent, so that it is evident that the clumping power does not reside in the fibrinogen alone. If, however, the

[1] The plasma keeps a part of its clumping power after filtration.

globulin is precipitated from *this* filtrate the latter is thereby deprived of its power to clump.

Devoto (*Cron. d. Clin. Med. d. Genoa*, 1896, Ann. 4, Part I.), after separating the globulins of typhoid blood by precipitation with ammonium sulphate, found that chemically they were identical with those of normal blood. The dried globulins retain their clumping power eight to ten days at least without loss, but are entirely without bactericidal power.

Whether the clumping material is simply carried down with the globulin and fibrinogen mechanically on precipitation, or whether it is more integrally connected with them, is not easy to decide. The results of experiments in dialysis do little to clear up the matter. No liquid possessing agglutinative power passes the membrane unaccompanied by albuminoid substances; that is, no clumping power appears in the dialyzed liquid until albuminoids begin to come through. On the other hand, agglutinative power is often present in non-albuminous *urine*, and some dialyzed sera which do contain albumin will *not* produce clumping.

On the whole, chemical researches have done very little to increase our understanding of the nature of the clumping power.

3. Facts have been adduced in former chapters which show the great *resisting power* of the agglutinating substance. It keeps for months in fluid serum *even if contaminated* by bacterial growths. Widal kept a specimen of serum fourteen months, at the end of which time its surface was thickly coated with moulds; yet its clumping power (1:16,000) was unimpaired. Sunlight has, according to Achard, no tendency to diminish the potency of typhoid serum.

As to the effects of heat, experiments made by Hayem (*Société médicale des Hôpitaux*, January 8th, 1897) show that a temperature of 60° C. does not deprive serum of its power to agglutinate.

Widal tested the milk of an immunized goat. Milk is not coagulated, as serum is, by heat, and therefore is better suited to experiments of this kind. He found that after a temperature of 66° C. was reached the clumping power of the milk began to diminish, and was altogether lost after ten minutes at 75° C. Later, the milk of the same animal was brought by four months' continuous inoculation to have a clumping power of 1:400. This milk stood ten minutes' exposure to a temperature of

75° C. without entirely losing its power. But when heated to 80° C. its power was entirely abolished.

Typhoid bacilli grown in bouillon from which a culture of colon bacilli has been removed by filtration are not deprived of their susceptibility to typhoid serum. Furthermore, the presence of other organisms (*e.g.*, colon bacilli) in a culture of Eberth's bacilli does not prevent a typhoid serum from finding out and clumping the typhoid bacilli.

Richardson has confirmed this observation, originally made by Widal and Sicard.

The Nature of the Clumping Process.

The fact that the discovery of the phenomenon of agglutination was made in the course of studies on immunity, together with the fact that the phenomenon is very marked with the blood of immunized animals, has led some observers to the assumption that the process of clumping is of itself in some way an expression of immunity.

This assumption is strengthened by the physical appearances of the process itself. When one watches the active typhoid bacilli become motionless and *apparently* lifeless under the influence of typhoid serum, and bunch themselves together "as if for mutual support and consolation," it is difficult to resist the impression that they are being acted upon by some force that is hostile to their existence. But, as has been repeatedly pointed out by Widal, Fraenkel, and others, the agglutinating and paralyzing powers of the serum are quite separate from its bactericidal or lysogenic powers. A sufficient amount of even normal serum will kill certain races of typhoid bacilli without giving rise to the slightest trace of clumping. On the other hand, clumped bacilli will grow readily upon ordinary media even if the clumping has been instantaneous and complete. Indeed, typhoid bacilli will multiply freely in (agglutinative) typhoid serum.

The production of Pfeiffer's phenomenon (lysogenesis) is rarely to be observed to any extent in agglutinated typhoid bacilli. At the centre of some large heaps we may notice a slight degree of granular transformation, but as a rule there is none; and in the peritoneal cavity of animals where Pfeiffer's test is best carried out, it is very difficult, as Salimbeni has shown, to produce any clumping. Air favors clumping, and the conditions existing within the animal organism do not. Exactly the reverse is true of lysogenesis. Animals often can be made immune without giving them clumping power, or can be given clumping power without any increase of immunity.

If these proofs were not enough, we have the support of clinical evidence, which makes it clear beyond all doubt that the power to clump is not an expression of immunity in any ordinary sense. Just when the agglutinating power is most intense the patient may testify to his lack of immunity by having a relapse. Or, again, the reaction may be very intense during the earliest days of the disease and gradually decrease as convalescence (and presumably immunity) advances.

Widal deserves all the credit for making this clear. He himself prefers to speak of agglutination as a phenomenon or reaction expressing "*infection*" rather than immunity. This term is good, negatively—*i.e.*, it prevents our falling into the mistake of associating the reaction with immunity. On the other hand, it seems insufficient to explain or express some of the facts regarding agglutination. For example, it is well known that the reaction may persist in the serum for periods of ten, twenty, even thirty years after the attack of fever which gave rise to it. Are we to consider the reaction as still a sign of infection thirty years after the patient has recovered? It is possible that a focus of infection may actually linger all these years in an otherwise healthy and effective organism. It may be that in all such cases bacilli remain in the gall-bladder, or in the urinary bladder, or in the bone marrow, as some recent investigations suggest, and respond for the whole organism as a healed tuberculous focus may respond to tuberculin. Further investigations along this line are most important.

But, again, if the reaction is one expressing infection, why is it not most marked in the worst cases or at the acme of a single case, and how are we to explain cases in which the reaction progressively increases in intensity as the patient's symptoms abate, or those cases in which the reaction first appears in convalescence?

Once more, how are we to explain the slight agglutinative powers not infrequently to be found in normal serum, or in the serum of persons suffering from diseases other than the one in question? We cannot say that there is a slight amount of infection in these cases and a larger amount in the disease itself.

PART II.

CHAPTER VI.

THE OCCURRENCE OF THE SERUM REACTION IN TYPHOID FEVER.

I HAVE collected in the tables on pages 61 to 63 5,978 cases of typhoid fever in which the serum reaction has been tested, and 5,668 cases of other diseases. It is very difficult to make such tables perfectly accurate, since the statements of the observers reporting the cases are not always perfectly clear; and, moreover, it is difficult to be sure that the same cases are not reported twice in different articles. I do not suppose that these tables include all the published cases, since many are not accessible to me, but I think the number collected is sufficient to make clear the general laws as to the occurrence of the serum reaction. On the whole there is a remarkable degree of unanimity among the opinions of the different observers all over the world. Out of more than one hundred who have expressed themselves on the subject all but three or four give practically the same testimony, and that testimony is identical with the conclusions to be drawn from the following tables.

OBSERVERS.	TYPHOID CASES.		CONTROL CASES.	
	Positive Reaction.	Negative Reaction.	Positive Reaction.	Negative Reaction
Wilson and Westbrook.............	687	0	8	268
Johnston.....................	128	1	0	400
Biggs and Park...................	180	60*		
Courmont......................	240	0	1	64
Cabot.........	491	4	1	625
Widal..........................	176	1	0	350
Shattuck.......................	122	1	3	101
Gehrmann and Wynkoop......... .	128	11*	1	50
Abbott.........................	858	—[1]	—[1]	699
Colville and Donnan.......	103	2*	1	19
Kerr...........................	120	0	1†	48

Cases marked with a star* were tested only once.
Cases marked with a dagger† may have had typhoid in the past.

[1] The total number of cases in which the clinical and laboratory diagnoses disagreed was 43 or 2.8 per cent.

OBSERVERS	TYPHOID CASES.		CONTROL CASES.	
	Positive Reaction.	Negative Reaction.	Positive Reaction.	Negative Reaction.
Thompson	157	6*	20	319
Fison	99	8*	0	59
MacKensie	57	4*	0	21
Levy and Gissler	105	0	0	10
Brill	160	2	1†	211
Barber	156	1	0	49
Musser	62	0	0	39
C. Fraenkel	66	0	0	"many"
Chantemesse	70	0		
Aaser	90	1	1	54
Elsberg	53	2	1	147
Knears	43	2*	5	38
Pakes	23	0	0	30
Robertson	19	0	1	25
Freyer	31	0	0	30
Anderson	27	2*	0	6
Paton	24	0	1	39
Weaver	27	3*	0	"many"
Herzog	15	0	1	several"
Villiès and Battle	31	0	1	52
Vanlair and Beco	16	0		
Ullman and Wöhnert	19	0	0	8
Uhlenhuth	15	0	0	16
Thiroloix	21	0		
Thomas	24	4*	3	18
Loomis	16		5	5†
Stern	101.	0		201
Reed	28	0	0	6
Pick	20	0	0	9
Brannan	19	1	1	79
Oordt	10	1	1	
Miller	28	3	97	8
Le Fevre	11	0	122	3
Kose	21	0	0	6
Jemma	12	0	0	15
Haedke	22	0	0	20
E. Fraenkel	14	0	0	1
Dempsey	14	0	0	4
Craig	9	0	0	12
Bartlett	12	2*	0	18
Ziemke	6	0	6	22
A. Fraenkel	5	0		
Pfuhl	8	0		
Stadelmann	4	0		
Whittaker	5	0		
Rochemont	5	0	0	17
Wright and Smith	16	3	0	15
Sabrazés and Hugon	13	1	0	9
Theolen and Mills	12	0		
Lemoine	4	0	0	7
Grünbaum	8	0	0	32
Durham	5	0		
Butters	4	0	0	0
Starck, Siegert, and Lambert—each 2	6	0		

Cases marked with a star* were tested only once.
Cases marked with a dagger† may have had typhoid in the past.

THE SERUM REACTION IN TYPHOID FEVER.

OBSERVERS.	TYPHOID CASES.		CONTROL CASES.	
	Positive Reaction.	Negative Reaction.	Positive Reaction.	Negative Reaction.
Ransom	4	0		
Brown	8	0	0	92
Comba	13	0		
Da Costa	95	7*	4	112
Appel and Thornbury	50	0	1	24
Greene	27	0	0	33
Block	43	3	3	?
Hausbalter	27	0	0	18
Gerloczy	37	0	0	50
Förster	26	0	1	89
Coleman	11	0	0	10
Catrin	86	0	0	21
Breuer	43	0	0	27
Bormans	20	0	0	0
Bensaude	83	0	1	141
Pugliesi	33	0	0	"many"
Gossage	21	0	2	141
Murray	34	0		
Hofmann	31	0	0	25
Mills	28	0		
Kühnan	9	0	0	50
H. Jackson	70	2*	1	26
TOTALS	5,814	164	323	5,345

Cases marked with a star* were tested only once.
Cases marked with a dagger† may have had typhoid in the past.

Out of 5,978 cases of typhoid here collected, 5,814, or 97.2 per cent, are stated to have shown a positive reaction, and only 164, or 2.8 per cent, a negative reaction; while out of 5,668 cases of diseases other than typhoid only 323, or 5 per cent, showed any serum reaction.

Despite certain sources of error, presently to be mentioned, these figures, I think, speak for themselves, and all the more eloquently because the results of different observations in various parts of Europe and America, and with all sorts of technique, are in the vast majority of instances practically identical.

Of course, some of the diagnoses may have been wrong, some of the sets of cases may have been selected ones, and different observers may mean different things by a "positive reaction" or a "negative reaction." Different men use different dilutions and different time limits, and some work with dried blood, some with fluid blood or serum, some by the microscopic and some by the macroscopic method. But after allowance has been made for all these differences, it seems to me that the statistics are essentially true and very important.

The majority of those who have used the method for two years, and have worked out the technique according to their personal equations, have learned to rely more and more upon the test. The number of negative tests in sure typhoid cases is, I am sure, larger than it would have been had the negative cases been retested. Most of the cases reported as negative were tested only once. As we shall presently see, it is not uncommon for the reaction to be late in making its appearance.

On the other hand, the positive tests in non-typhoid cases were very few of them made under such conditions that typhoid could be certainly excluded. I shall return to this subject later.

How Early Does the Reaction Appear?

It is very difficult to get satisfactory evidence upon this question, owing to the carelessness of most writers as regards the manner of dating the beginning of the disease. The great majority of articles on serum diagnosis contain some statements as to the earliest day of the disease at which the reaction appeared. But it is hardly ever made clear *what is meant by the first day of the disease*. Some reckon from the earliest symptom of any kind, including the more or less indefinite prodromata as a part of the course of the disease itself. Others date from the occurrence of some special symptom like headache or epistaxis. Others reckon the day on which the patient feels sick enough to give up work as the first day of the disease, while still others date the beginning of the infection from the first day in bed. It is obvious how confusing statistics upon the subject must be.

Another point of obscurity in the majority of articles is the expression "The earliest reaction occurred on the —th day," not specifying whether the blood had been tested previous to this time, or whether the writer merely means that the reaction was present, at least, as early as the —th day, and very possibly earlier.

Thus Widal[1] says that he has six times succeeded in getting a positive reaction on "the fifth day," but he does not state how he reckons the beginning of the disease, nor whether he had tested these six cases on previous days.

On the whole, perhaps, the most definite information that

---

[1] Widal: Annales de l'Institut Pasteur, May 25th, 1897.

we have on the subject is that afforded by inoculation experiments in animals. The blood of guinea-pigs inoculated subcutaneously with 1 to 1½ c.c. of bouillon culture of Eberth's bacilli usually begins to present agglutinative power on the third, fourth, or fifth day after the injection. In a guinea-pig inoculated by Widal with 1 c.c. of a twenty-four-hour bouillon culture the reaction first appeared sixty hours, or two and one-half days, later with a potency of 1:15. Later the potency gradually and spontaneously increased up to 1:500.

If cultures are heated to 60° C. for forty-five minutes it needs 2 c.c. to bring about an agglutinating power in the blood, and this does not appear for five days. Boiled cultures are effective only in doses of from 10 to 12 c.c.

Cultures filtered through porcelain without being exposed to heat or chemical antiseptics were found still capable of producing the agglutinating power in a guinea-pig's blood, but only after waiting eight days and using a dose of 8 c.c.

Widal also experimented with the serum of an ass which had a clumping power of 1:30,000. Two guinea-pigs were inoculated with this serum, one receiving twelve drops subcutaneously, and the other twelve drops in the peritoneal cavity. Each pig began to show agglutinative power in the blood in one-half an hour (1:10). The potency of the first pig's serum reached its maximum in ten hours, attaining 1:180. The second reached 1:250 in ten hours. In the course of ten days this agglutinative power gradually disappeared.

Park (New York *Medical Journal*, March 27th, 1897) got no reaction in guinea-pig's blood till the sixth day after inoculation with live bacilli.

The figures published by Bensaude (*loc. cit.*, p. 74) are among the clearest statements that we have as to the first appearance of the reaction in the typhoid infections of man. He makes it clear in the first place just what he means by the first day of the disease: namely, *the first day in bed*. Out of 45 cases in which the reaction was present at the time of entering the hospital, he finds—

3 cases at least as early as the 3d day in bed—possibly earlier.
2 "   "   "   "   "   4th "   "   "   "
6 "   "   "   "   "   5th "   "   "   "
5 "   "   "   "   "   6th "   "   "   "
6 "   "   "   "   "   7th "   "   "   "

7 cases at least as early as the 8th day in bed—possibly earlier.
2 "   "   "   "   9th "   "   "   "
1 case   "   "   "   10th "   "   "   "
1 "   "   "   "   11th "   "   "   "
4 cases "   "   "   12th "   "   "   "
3 "   "   "   "   13th "   "   "   "
3 "   "   "   "   15th "   "   "   "
1 case   "   "   "   17th "   "   "   "
1 "   "   "   "   24th "   "   "   "

In 7 other cases the reaction was not present at the time of entrance, and appeared *only* between the following dates:

4th to 6th day............................ .... 1 case.
4th to 11th day .... 1 "
6th to 8th day................................. . 1 "
6th to 11th day... .... 1 "
8th to 10th day .... 1 "
18th to 39th day................................ 1 "
16th day to 1 week after convalescence .... 1 "

Summing up these results we find that in 26 cases out of 52 (or just one-half) the reaction was found before the eighth day in bed. In 9 more of these cases it *may* have been present on the eighth day or earlier, but was not tested for. In only 3 cases was it *proved* that the reaction was not present before the eighth day in bed. In only 2 cases was the reaction known to be delayed in appearing beyond the fourteenth day.

Villiès and Battle (*Archives Générales de Médecine*, July, 1897) among 24 cases which showed reaction when first examined found the following: 2 cases were on the "fourth day of the disease" (query: how reckoned?); 6 on the fifth day; 3 on the sixth; 3 on the seventh; 1 on the eighth; 2 on the ninth; 2 on the fourteenth; 1 on the fifteenth; 1 on the seventeenth; 1 on the twenty-fourth; 2 on the thirtieth.

In one case the reaction was absent on the third day, but present on the fifth. In another, the reaction was absent on the sixth day, but present on the ninth.

Out of 26 cases 15 are known to have reacted before the eighth day, and only 1 is positively known to have had no reaction until the ninth. The others were not tested till after the eighth day, and nothing can be said, therefore, as to the date at which they first showed the reaction.

G. B. Shattuck (*Medical News*, May 8th, 1897) gives the following figures: There was reaction on the sixth day in 1 case; on the seventh day in 4 cases; on the eighth day in 6 cases; on the ninth day in 5 cases; from the tenth to fifteenth day in 23 cases; from the fifteenth to the twentieth day in 16 cases; from the twentieth to the thirtieth day in 32 cases; from the thirtieth to the eightieth day in 33 cases. Total number of cases, 120.

De Rochemont (*Münch. medicinische Wochenschrift*, 1897, No. 5) got a positive reaction in one case on the third day from the first symptoms (1:40) and one on the fourth day from the earliest symptom (1:100).

Johnston and McTaggart (*British Medical Journal*, December 5th, 1896) report cases positive on the second day.

In Elsberg's article (New York *Medical Record*, April 10th, 1897) the following figures are given. (The beginning of the disease is apparently reckoned from the first day in bed.)

Positive reaction on 4th day in 1 case.
"   "   " 5th day in 2 cases.
"   "   from 7th to 12 day in 20 cases.
"   "   " 12th to 14th day in 8 cases.
"   "   " 14th to 17th day in 2 cases.
"   "   " 24th to 37th day in 3 cases.
    Total number of cases, 36.

He considers that the reaction is present within the first week in 8 per cent of cases, in the second week in 78 per cent, and in third or later in 14 per cent.

Courmont (*La Semaine médicale*, 1897, p. 209) found reactions before the sixth day in 116 out of 240 cases, or nearly 50 per cent, and reactions by the ninth day in 235, or 93 per cent.

Catrin (*ibidem*, 1896, p. 418) tested:

4 cases on the 3d day and found them all negative.
6   "   "   4th   "   "   4 positive and 2 negative.
1 case   "   5th   "   "   it negative.
3 cases "   6th   "   "   all positive.
1 case   "   7th   "   "   it positive.
4 cases "   8th   "   "   all positive.

Biggs and Park (*American Journal of the Medical Sciences*, February, 1897) report the results of the first examination of 108 cases:

Of 19 cases tested between 3d and 7th day, 63 per cent were positive.
" 39 "   "   "   7th " 14th " 59 "   "   "
" 24 "   "   "   14th " 21st " 79 "   "   "
" 9 "   "   "   21st " 28th " 88.8 "   "   "
" 17 "   "   "   30th " 60th " 76 "   "   "

Da Costa (New York *Medical Journal*, August 21st, 1897) summarizes his results in the following table:

Day of Disease.	No. of Cases.	Positive.	Negative.	Per cent Positive.
4th to 7th	14	13	1	92.8
8th to 14th	39	36	3	92.3
15th to 21st	24	24	0	100
22d to 28th	8	8	0	100
29th to 35th	8	7	1	87
36th to 42d	5	4	1	80
43d	1	1	0	100
50th	1	1	0	100
51st	2	1	1	50
	102	95	7	

Barber (New York *Medical Journal*, 1898, No. 16) gives the following figures:

```
 4 cases showed Widal's reaction on the  1st day, and  1 case had diazo on that day.
13    "    "    "    "    "    "         2d   "    "   15 cases "    "    "    "    "
12    "    "    "    "    "    "         3d   "    "   21   "    "    "    "    "
25    "    "    "    "    "    "         4th  "    "   22   "    "    "    "    "
20    "    "    "    "    "    "         5th  "    "   18   "    "    "    "    "
 8    "    "    "    "    "    "         6th  "    "   18   "    "    "    "    "
22    "    "    "    "    "    "         7th  "    "   16   "    "    "    "    "
 7    "    "    "    "    "    "         8th  "    "    6   "    "    "    "    "
 3    "    "    "    "    "    "         9th  "
11    "    "    "    "    "    "        10th  "    "    4   "    "    "    "    "
 2    "    "    "    "    "    "        12th  "    "    3   "    "    "    "    "
 3    "    "    "    "    "    "        14th
 2    "    "    "    "    "    "        15th
 1 case  "    "    "    "    "          21st  "   Negative on 6th and 10th days.
 1    "    "    "    "    "    "         23d  "       "       " 21st day.
 1    "    "    "    "    "    "         28th "   First examination.
 1    "    "    "    "    "    "         60th "   Negative on the 6th day.
 1    "    "    "    "    "    "         75th "       "       " 3d and 5th days.
 1    "    "    "    "    "    "        113th "       "       " 7th day.
138 cases.
```

Bracken (Philadelphia *Medical Journal*, Vol. I., No. 2, 1898) reports positive agglutinative reaction in the blood: On the first day, 3 cases; second day, 26 cases; third day, 29 cases; fourth day, 53 cases; fifth day, 38 cases; sixth day, 44 cases; seventh day, 70 cases. Total, 263 cases.

It was absent until the 8th day in 7 cases (later positive).
                      9th   " 5 "   "
                     10th " 5 "   "
                     12th " 8 "   "
                     14th " 3 "   "
                     22d   " 1 case "
                     36th " 1 "   "
                     40th " 1 "   "
                                  31 cases.

Thus in 263 out of 294, or 89 per cent, the reaction was present during the first week.

Out of 13 cases examined in the first week of the disease by Levy and Gissler[1] 10 showed positive reaction, and the others reacted later.

Out of 22 cases in the 2d week 22
       24   "   3d   " 24
       16   "   4th " 16
       13   "   5th " 13
        7   "   6th "   7 } showed positive reaction.
       10   "   7th " 10
        5   "   8th "   5
      97 cases.

Barber has some remarkable records of early reactions. For example, in a child of three and one-half years who had been recognized as sick only eleven hours before the test, the reaction was positive, and the case proved to be one of typhoid. Barber got a positive reaction also in an adult male whose symptoms dated back only twenty-four hours.

Comparing the Widal test with the diazo reaction, he finds the former by far the more reliable of the two, since it was present in 155 out of 156 cases, while the diazo reaction was found in only 128 out of 137 cases. He believes, however, that the diazo appears usually a little earlier than the agglutinative reaction.

In one case he found that the serum test was present on the tenth day, absent on the seventh, fifteenth, and eighteenth days, present again on the twenty-third, absent on the twenty-sixth. Such variations in the condition of the blood have been noticed

---

[1] Levy and Gissler: Münch. med. Woch., December 14th, 1897.

by all who have made daily tests in any considerable number of cases. (Barber used dried blood with a dilution of 1:25.)

Elsberg (*loc. cit.*) has called especial attention to this point. Testing cases daily he found that the reaction appeared *suddenly*—i.e., no trace of agglutination on one day, and a marked clumping the next. This, he says, is the *rule*. Occasionally it may come on gradually.

Brill reports on 160 cases, 158 of which showed positive reactions. Of 80 of these seen at the Mount Sinai Hospital, 2 reacted positively on the fourth day, 4 on the fifth day, 44 between the seventh and twelfth days, 12 between the twelfth and fourteenth days, 10 between the fourteenth and seventeenth days, 6 between the twenty-fourth and thirty-second days.

In the late cases no early tests were made, so that the reaction may have been present earlier.

Aaser[1] reports: Reaction as early as the third day in 2 cases; fourth day in 4 cases; fifth day in 2 cases; sixth day in 3 cases; seventh day in 4 cases.

Reaction in the second week in 24 cases; third week in 10 cases; fifth week in 8 cases; sixth week in 19 cases; seventh week in 8 cases; ninth week in 6 cases.

I am not sure whether any of the late cases were tested then for the first time or not.

Henry Jackson (personal communication) gives these figures, each of which represents the first test made in any case.

Reaction at least as early as the sixth day in 9 cases; tenth day in 21 cases; fifteenth day in 16 cases; second to fourth week in 12 cases; third day in bed in 1 case; seventh day in bed in 7 cases; fourteenth day in bed in 4 cases. Total, 70 cases.

Many of the cases in my series I did not see till they were well advanced in the disease, either because they entered the hospital late, or because I could not get time to examine them. Making allowance for all these drawbacks my results are as follows:

Out of 35 cases in which the reaction was tried within two weeks from the time when the patient gave up and went to bed it was present:

[1] Reference in Journal of the American Medical Association, April 24th, 1897.

On the 1st day in bed or earlier in 2 cases.
    2d   "   "   3 "
    3d   "   "   1 case.
    4th "   "   5 cases.
    5th "   "   2 "
    6th "   "   1 case.
    7th "   "   6 cases.
    8th "   "   3 "
    9th "   "   4 "
   10th "   "   1 case.
   11th "   "   0 "
   12th "   "   1 "
   13th "   "   4 cases.
   14th "   "   2 "

        Total .... 35 cases.

Again, out of 18 cases in which the blood was examined within two weeks from the first day on which the patient felt in any way under the weather, the reaction was present:

On the 5th day of "malaise" in 1 case.
    7th "   "   2 cases.
    8th "   "   1 case.
    9th "   "   5 cases.
   10th "   "   2 "
   11th "   "   2 "
   12th "   "   1 case.
   13th "   "   1 "
   14th "   "   3 cases.

        Total .... 18 cases.

In both these series of cases, as in most others, the figures are somewhat misleading, as in no case was the blood examined at all *before* the day on which positive results were obtained, so that in every case the reaction was very possibly present *before* it was tested for. All that the figures show is that the serum reacts at least *as early as the fifth day* of malaise and the first day in bed, and *very likely earlier*.

As regards the relation of the date of appearance of the serum reaction to that of the diazo reaction, rose spots, and splenic enlargement, the following cases are of interest:

In two cases the reaction was noticed five days before the appearance of rose spots or splenic enlargement; once it anticipated the diazo reaction by two days.

In seven cases in which neither rose spots nor splenic enlargement were ever present, the serum reaction was characteristic of typhoid; it was also positive in two cases in which diazo reaction was absent throughout.

These points were not specially attended to, and show merely that the serum reaction *sometimes* appears before the diazo reaction, splenic tumor, or rose spots, without giving us any idea of how often this is the case.

Abbott (*Medical News*, May 8th, 1897) has twenty-three times obtained the reaction within the first week of the disease. C. Fraenkel, Johnston, and Sabrazés and Hugon mention positive reactions on the second day; Musser, Pick, Craig, Appell and Thornburg, and Villiès and Battle on the third day (the latter observers found it twice on the third day); Ullman and Wöhnert and Chantemesse and Thiroloix on the fourth day; LeFevre, Pick, and Craig on the fifth day; Breuer, LeFevre, Osler, and Ullman on the sixth; Ziemke, Musser, and Oordt on the seventh; Loomis, Menetrier, and Délepine on the eighth; Stern on the ninth.

Thomas (New York *Medical News*, 1898, No. 18) reports that in sixty per cent of his cases the reaction was present before the twelfth day.

Summary.

It is difficult to generalize from these figures owing to the difficulties (above mentioned) of finding out what is meant by the first day and the scantiness of reports as to the results of tests previous to the day on which the reaction was positive.

Out of a total of 849 tested before the eighth day there were 93 per cent positive reactions. Courmount's figures are the most intelligible, and according to him at least 93 per cent of all cases react within the first nine days.

How Late may the Reaction Appear for the First Time?

Wilson reports 1 case in which the reaction was absent as late as the thirty-sixth day; Pick reports 1 case in which the reaction first appeared on the thirty-fourth day; Kolle cases in which the reaction first appeared on the sixteenth and seventeenth days; Durham 1 on the eighteenth day; Stern 1 at the end of the second week; Widal 1 on the twenty-second day;

Breurer 1 in the eighth week; Thoinot 1 in the eighth week; Bensaude 2 cases on the eighteenth and twenty-fifth days; Achard 1 on the fourth day of apyrexia; Castaigne 1 case on the fourth day of apyrexia; Biggs and Park 1 case in the fourth week, and 1 case in relapse; Abbott 1 case on the thirty-second day; Elsberg 4 cases on the sixteenth, twenty-fourth, twenty-seventh, and thirty-sixth days; Barber 1 case on the thirty-third day (negative on thirtieth) and 1 case on the thirty-fifth day (negative on thirtieth); Blumenthal 1 case[1] on the twenty-ninth day (having been negative on the twelfth and twenty-first days); Gruber 3 cases on the twenty-fifth, thirtieth, and thirty-fifth days (1 : 32).

Widal, Eshner, Breuer, Wilson, Cahill, and Thoinot and Cavasse have found the reaction absent until relapse; the last-named observers found no reaction in their case until the *second relapse*. Biggs and Park in 2 cases, and the author in 3 cases have also found the reaction absent until relapse.

In my series there were 19 cases out of 292 in which the reaction did not appear at all until after the second week in bed, and out of these 19, 10 did not react until the relapse or until the temperature had reached normal. The results of other observers seem to be approximately the same. Johnston finds "late or absent reactions" in only 1 to 2 per cent. Wilson and Westbrook in only 3 out of 294 cases found the reaction absent as late as the fourteenth day. Courmont found only 9 per cent of 240 cases in which the reaction was delayed as late as the tenth day.

How Early may the Reaction Disappear?

Widal has himself called attention to the fact that the reaction, so far from lasting after convalescence, may in certain cases disappear very soon after the temperature reaches normal. Thus Widal saw the reaction disappear on the eighteenth afebrile day in one case and on the twenty-fourth afebrile day in another. Breuer has found it gone as early as the seventeenth afebrile day in one case and the twenty-fifth in another. Thiercelin and Lenoble missed it after the twentieth afebrile day, Eug. Fraenkel on the twenty-fifth, while C. Fraenkel in three

---

[1] In this case typhoid bacilli were obtained from the blood by culture. Hence (according to Blumenthal) the absence of reaction.

cases occurring in children observed the cessation of the agglutinative reaction a few days after the fall of temperature. These last were light cases.

Elsberg records cases in one of which the reaction lasted only eight days in all, in another only twelve days. These were cases in which he was able to witness both the time of the reaction's appearance and the time of its disappearance, so that there could be no mistake about them. In such cases it is obvious how easily mistakes might occur and negative results be reported in case the observer does not happen to hit on the period during which alone the reaction exists. Fortunately such cases are very rare.

In one of Elsberg's cases the reaction was gone at the end of the fifth week, but in the vast majority it persisted several months.

In Bensaude's experience the earliest dates for the disappearance of the reaction were the tenth afebrile day in 1 case; the sixteenth in 1 case; the twenty-first in 1 case; the forty-second in 1 case; the forty-fifth in 1 case.

Two of these cases were *very* mild—what the French call "typhoïdettes;" two others mild, and the other of moderate intensity, and Bensaude accordingly concludes that light cases are more apt to show the reaction for short periods.

The question, "*Why does the reaction appear early in some cases and late in others?*" has been discussed by Widal and others. Certainly the earliest reactions do not come in the severest cases, nor in the mildest. Thoinot and Cavasse have supposed that a late reaction means a mild case, but Achard and Bensaude have reported the late appearance of the reaction in very severe cases, and I have myself seen a case die of perforation before the reaction had appeared at all. That the presence of typhoid bacilli in the blood is associated with some cases in which the reaction is late or altogether absent has already been mentioned, but this is not the fact in every case. No wholly satisfactory theory has yet been advanced.

The case of Blumenthal, above mentioned, and two cases reported by Anthony and Ferré (*Journ. de Méd. de Bordeaux*, 1897, No. 30), in which an absence of the agglutinative reaction was associated with the presence of live typhoid bacilli in the blood, certainly suggest, when taken in connection with Courmont's researches, that the presence of bacilli in the blood is inimical to the development of the agglutinating power.

The same thing occurred to me in connection with a case to be more fully described in chapter VIII., in which a very late and feeble serum reaction in a case of undoubted typhoid was associated with signs and symptoms suggestive of metastases.

Grünbaum states that "it is known that the agglutinating power is used up during its action on the bacilli," and thinks that this "explains why the most severe cases of typhoid may have comparatively slight agglutinating power in the blood" (*Science Progress*, October, 1897).

How Long may the Reaction Last?

Widal examined 40 cases in which a history of typhoid at least one year before had been obtained; 11 of these cases showed a positive reaction.

```
1 case   8 years after typhoid infection showed clumping at 1 : 1800
1  "     7   "          "         "        "         "     1 : 150
2 cases  9   "          "         "        "         "     1 : 40 and 1 : 30
1 case  26   "          "         "        "         "     at 1 : 30
```

Biggs and Park examined 12 cases with these results:

```
8 cases 3 to 4 months after typhoid infection— { 2 moderate
                                                 3 marked    } reaction.
                                                 3 very slight
1 case 1 year after typhoid infection—very slight.
1  "   3.years    "         "         0
1  "   5   "      "         "         very slight.
1  "  14   "      "         "         0
```

Elsberg tested 6 cases, one-half year to ten years after typhoid infection and found them all negative. Pick had similar results in 7 late cases (one to sixteen years after infection). Feson (*British Medical Journal*, July 31st, 1897) tried 21 persons at periods of from three months to eight years after infection and got reaction in 18, one of whom had had the fever seven and a half years before. Grünbaum got positive reaction in 6 cases after from four to thirty-seven (!) years from the time of infection. Musser and Swan tested 13 cases and found positive reactions in 2 of them four years and ten years respectively after infection. Villiès and Battle got positive reactions in 5 convalescents three, six, seven, twelve, and thirty months after infection. Rochemont tested 2 cases and Wright and Smith tested 17 cases with negative results.

Brannan out of 5 cases examined got clumping in 3 (at ten months after infection in 1 case and two months in the others). J. L. Miller found positive reactions in 2 out of 7 cured cases. These reactions were respectively four and twenty years after infection. Gehrmann (Chicago *Medical Recorder*, May, 1897) reports 17 cases in which he obtained positive reactions at periods of from three months to twenty-two years after typhoid infection. In 2 cases in which there was a history of typhoid ten and sixteen years before, the blood was negative. C. Fraenkel reports a positive reaction (1 : 20) in the blood of one of his students who had had typhoid thirteen years before. Courmont (*La Semaine médicale*, 1897, p. 207) states that in children the reaction is usually gone within two months. In adults it often lasts four or five months. These statements are based on an examination of 58 convalescent cases. Among 14 who had had typhoid one year or more before he got only 2 positive reactions. Uhlenhuth (*Deut. Militär-Zeit.*, 1897, Heft 3) reports a positive reaction eight years after infection. J. B. Thomas (*Medical News*, April 3d, 1897) reports reactions after three and six years, and J. S. Tew (*Lancet*, 1898, Vol. I., No. xii.) after four and nine years.

The most thorough study of late reactions with which I am acquainted is that made by E. Renard (*Thèse de Paris*, 1897). He tested 104 cases and got a reaction in 35. This includes positive reactions:

At 23 years after infection in 1 case.
" 24 " " 1 "
" 26 " " 1 "
" 27 " " 1 "
" 30 " " 1 "

All these cases were tested at a dilution of 1 : 10, except one case in which the reaction was still present at 1 : 40 a year after infection.

Widal measured the amount of clumping power present from time to time in the blood of two individuals who had had typhoid fever eight and nine years before. He found that the potency of the serum did not vary widely from day to day, as it does during the fever, but remained practically stationary.

Summary.

Putting all these reports together we have:

1 case as late as 37 years after infection.
1 "   "   30   "   "
1 "   "   27   "   "
2 cases "   26   "   "
1 case   "   24   "   "
1 "   "   23   "   "
1 "   "   22   "   "
1 "   "   20   "   "
1 "   "   16   "   "
2 cases "   14   "   "
1 case   "   13   "   "
1 "   "   10   "   "
3 cases "   9   "   "
3 "   "   8   "   "

And many after shorter periods.

Theory of Late Reactions.

A possible explanation for these late reactions is suggested by such cases as that lately reported by G. B. Miller (*Johns Hopkins Hospital Bulletin*, 1898, Vol. IX., No. 86). The patient was operated on for gall-stones *seven* years after an attack of typhoid fever. Live typhoid bacilli were found in the gall-bladder, and the patient's own serum reacted at 1:100. Recent observations by M. W. Richardson and others show that the urinary bladder may also serve as a repository for Eberth's bacilli long after the patient has recovered from his acute symptoms. The bones form another hiding-place in which typhoid bacilli may be stowed away for months or years. Thayer (*Johns Hopkins Bulletin*, March, 1897) mentions a case in which the persistence of the serum reaction sixteen months after convalescence may have been due to the existence of typhoid bacilli in a focus of osteomyelitis which was then present. Elsberg and Brannan (*loc. cit.*) report similar cases, and Park mentions a case of abscess of the liver one year after an attack of typhoid, which abscess was believed by him to contain Eberth's bacilli and to be the cause of the marked serum reaction which still persisted. But Da Costa, who examined the blood of a young girl having a suppurative arthritis of the knee due to the typhoid bacillus, did *not* obtain a decided positive reaction, although the patient recovered from a severe attack of enteric fever six weeks before.

CHAPTER VII.

THE SERUM REACTION IN TYPHOID FEVER (*Continued*).

INTERMITTENCE OF THE REACTION.

WIDAL early in his investigations noticed that the reaction sometimes is present one day, absent the next, and then again present on the third, and this observation has been confirmed by all who have looked into the matter.

For example Wilson (*loc. cit.*) quotes the following cases:

CASE A: Reaction positive on the 19th, 20th, 26th, 29th, and 34th days; negative on 9th, 13th, and 21st days.

CASE B: Reaction positive on 3d, 13th, 18th, 26th, 40th, 90th, and 123d days; negative on 25th day.

CASE C: Reaction positive on 2d, 5th, 10th, 52d, and 66th days; negative on the 21st day.

Elsberg (*loc. cit.*) has had similar experience.

RELATION OF AGGLUTINATION TO THE HEIGHT OF THE TEMPERATURE IN TYPHOID CASES.

Jemma, of Genoa, has made a careful study of the relation existing between the intensity of the agglutinating power of the serum and the temperature of the patient. He finds that as a rule the clumping power is greatest when fever is highest—*i.e.*, during the latter part of the day, despite the fact that if the serum is exposed to artificial temperature, such as 60° C., its power to clump is much diminished.

THE CLUMP REACTION IN CASES VACCINATED AGAINST TYPHOID.

J. S. Tew in a recent paper (*Lancet*, 1898, Vol. I., No. 12) reports examination of the blood in 21 cases which had been vaccinated against typhoid fever, using injections of dead cultures. In 18 of these, or six-sevenths of all, the clump reaction had appeared in the serum within a few days.

Wright found the reaction on the second day after injection (*British Medical Journal*, 1897, p. 156) in 18 cases, the agglutinating power often reacting 1:50.

C. Fraenkel has suggested that in some cases there may be a natural immunity against typhoid, as seems to be the case with cholera, yellow fever, and diphtheria. He instances the case of a little boy who was the only one of a large family to escape an attack of typhoid fever during a recent epidemic. This boy's blood showed an agglutinating power within two hours at a dilution of 1:50. The facts might be otherwise explained, however. All the weight of evidence goes to show that the reaction is one of infection rather than of immunity, and it is possible that the boy had typhoid bacilli enough in his gall-bladder or elsewhere to produce a clumping power in his blood without experiencing any subjective discomforts. Such a case has been reported by Cushing (*Johns Hopkins Bulletin*, 1898). Live typhoid bacilli were found at operation in the gall-bladder. No history of typhoid could be obtained.

Value of Serum Diagnosis in Typhoid Fever.

I think we can best sum up the evidence hitherto presented by attempting to answer the following questions.

I. What inference should be drawn from a negative serum reaction in a case of suspected typhoid?

II. Does a positive serum reaction always mean typhoid?

(a) There is no reason to doubt that the agglutinative reaction is occasionally though very rarely absent in cases of typhoid infection. Widal has himself published notes of a case in which typhoid bacilli were obtained from the spleen twice, first during the original attack and later in the relapse, yet even by Widal's own hands no serum reaction could be obtained at any period of the disease.

Stadelmann has seen one case with autopsy in which the serum reaction was equally negative, and I have had a like experience. In my case the patient died of perforation eight days after entering the hospital. His blood was examined three days before death and no reaction found, but autopsy showed typical typhoid lesions and cultures from the spleen were positive. In the other two cases in which I have failed to find any agglutinative reaction despite repeated trials, there was no autopsy, but clinically the picture was typical of typhoid.'

In three other cases I have failed to obtain a reaction at the first examination, and have subsequently had no opportunity to repeat the test. These cases I have not counted as negative on account of the absence of a second test. In most of the reports the negative cases include many which have not been retested, and some which would very possibly have shown a positive reaction later on, so that in the statistics collected on page 63 the number in the negative column is doubtless too large.

But even after all due allowance is made for late reactions, miscalled negative, for mistakes in diagnosis and in technique, the mistakes in the positive column are probably numerous enough to balance the error, and I have no doubt that in at least two per cent of cases of genuine typhoid infection there is no serum reaction to be obtained at any time.

Obviously, then, the absence of a serum reaction is not a sufficient ground for the inference that no typhoid fever exists, although if the test is *repeatedly* negative at various stages of the disease the chances are 980 out of 1,000 that the disease is not typhoid.

(*b*) But it is of the greatest importance to bear in mind that a negative reaction in the earliest days of the disease affords only a presumption against the existence of the disease. This presumption becomes the stronger the later in the disease the reaction is tested. Bensaude found only 2.4 per cent of "late reactions" in his 83 cases. Widal mentions only 3 late reactions in his 177 cases, or 1.7 per cent. I have seen 21 cases out of 491, or 4.2 per cent, in which there was no reaction until after the second week, and among these, 10, or 2 per cent, showed no clumping power until after the temperature had reached normal. Courmont's figures showed 9 per cent of cases (out of 240) in which the serum reaction was delayed until the eleventh day or later. Johnston finds in only 1 or 2 per cent of his cases an "absent, feeble, or delayed reaction."

(*c*) The intermittence of the reaction from day to day furnishes another source of error in case we fail to retest our cases.

An absence of serum reaction in a case supposed to be typhoid may mean then: (*a*) that the reaction is delayed and will appear later; (*b*) that the reaction has already appeared

[1] These three negitive cases are the only ones in my series of 491 cases.

and disappeared again for good; (c) that the reaction has happened to intermit on the day of the test; (d) that there is no reaction to be obtained in this case, though typhoid infection exists; lastly (e) that the reaction is absent because no typhoid infection exists. This last is the proper inference in ninety-eight per cent of all cases *provided* we have excluded (a), (b), and (c), as we easily can in most instances.

The fact that the reaction is usually absent during the earliest days of the disease has led many to the conclusion that its value is slight, since it is only in the earliest days of the disease that the diagnosis is difficult. But in answer to this it should be said:

1. That the reaction usually *is* present when the patient first consults a physician, though not when he first begins to feel poorly. In most parts of the country typhoid is chiefly a disease of the less educated classes, and hence little attention is paid to the earliest symptoms, and almost all cases are "walking typhoids," in the sense that they keep about for a day or two after they begin to have fever. Over ninety per cent of the cases seen at the Massachusetts General Hospital since 1896 have shown a typical reaction at the time of entrance.

2. But even when the reaction does not appear until the second week or later, it is of the greatest diagnostic value in many cases—especially: (1) in the diagnosis of abortive or atypical cases; (2) in the differential diagnosis between typhoid and typhus, malarial, remittent, dengue, Malta fever, yellow fever, and other infections of tropical or semitropical countries; (3) in the identification of obscure forms of tuberculosis or deep-seated suppurative processes; (4) in the differentiation of typhoid from pneumonia and influenza; (5) in the retrospective diagnosis of typhoid.

This last application of the serum test is of value not only for diagnosis, prognosis, and treatment (as will be exemplified presently), but in the investigation of epidemics and from various medico-legal standpoints. Examples of these will be given later.

Agglutination of Typhoid Bacilli by Sera Other than those from Cases of Typhoid Fever.

1. It has been repeatedly stated that the phenomenon of clumped typhoid bacilli does not mean that the clumping has

been caused by the blood of a case of typhoid fever. Under certain conditions typhoid bacilli clump without the addition of any serum whatever. In old bouillon cultures left in the thermostat at 37° C., or even at room temperature in bouillon not properly prepared, and sometimes under conditions not easily identified and without known cause, spontaneous clumping may occur. If one happened to add normal serum to such culture and then examined the mixture and saw the clumps, one might naturally though illogically conclude that the clumping was due to the serum. I have no question that a large number of the cases reported as positive reactions with non-typhoid serum are to be explained in this way.

2. Other cultures which have no tendency to spontaneous agglutination will undoubtedly show agglutination with non-typhoid serum, provided certain precautions as to dilution and time limit are disregarded. When mixed with serum in equal proportions, instead of 1:10, or when left in the thermostat for an hour or more in contact with serum from non-typhoid cases, some cultures will certainly show clumping. Many errors doubtless arise in this way. But this is only admitting that if wrongly performed the test may lead us to erroneous conclusions.

3. Many persons stated to have been free from typhoid fever may really have passed through an attack under another name, or even without knowing that they were ill at all. It is not at all infrequent to see a case of typhoid fever in what we have good reason to believe is a relapse, although the patient was up and about his business a few days before. If the relapse had never occurred, that patient might never have known that he had passed through typhoid, and might so have informed an investigator who desired to use him as a "control" years after.[1]

4. It can hardly fail to strike any one who looks through the literature of "typhoid reactions in non-typhoid cases" that no two observers find these pseudo-reactions in the same diseases. Almost as many diseases are represented as there are observers, and as often as not it is with normal blood that these pseudo-reactions appear. This seems to me to suggest very strongly that such pseudo-reactions are due to conditions existing not in the blood of the disease in question, but in the technique. If any non-typhoid blood had anything like the clumping power contained in typhoid blood, we should find different observers

[1] Compare the case quoted from Cushing on page 79.

specifying the same disease or diseases again and again. As it is, the finding of a single positive clump reaction by one observer, say in diabetic coma, by a second in pernicious anæmia, by a third in malaria, by a fourth in the blood of a healthy negro, and so on, seems to me to mean that about once in so often a technical error is apt to creep in, which error happens to come now when this disease and now when that is being tested.

In some of the published cases this is obviously the explanation. Thus in the case of puerperal septicæmia (with autopsy) reported by Ferrand and Theoari (*La Semaine médicale*, 1897, p. 30) the twenty-four-hour macroscopic test was made, as Theoari himself admitted, with impure cultures. In the microscopic test it is easy to suppose that spontaneous clumping had taken place, since no previous scrutiny of a drop of the pure culture is reported.

Stern (*Berliner klinische Wochenschrift*, 1897, No. 11), who got a positive agglutinative reaction in 20 out of 70 non-typhoid blood, uses a technique well calculated to lead to spontaneous clumping without the addition of any serum whatever. He leaves his slide and cover preparations for two hours in the thermostat at 37° C. This accounts for the great difference between his results and those of any other observer, and it is no wonder that he discards the 1:10 dilution and insists that it should be at least 1:50, for his technique so magnifies any natural tendency to clumping which may exist that he has to take special precautions to reduce it again. The heat and dryness of the thermostat hasten and promote clumping unnaturally. This accounts for Stern's extraordinary report of clumping as high as 1:30 in two cases and 1:20 in five cases, all in non-typhoid blood.

To compare these results with those obtained with a 1:10 dilution and a fifteen-minute time limit at room temperature, and to suppose the two contradictory is illogical. Results obtained in such different ways cannot be held to contradict.

Stern's results are, as Widal says, a useful warning against the use of long time limits at high temperature.

Other reports of positive serum reactions in "non-typhoid" cases are to be explained by the want of thoroughness in the post-mortem examination. For example, in the much-quoted case of Jez[1] a positive clump reaction was obtained during life

[1] Jez: Wien. med. Woch., January 16th, 1897.

and at autopsy tubercular meningitis was found. No cultures were taken from the organs. Who can be sure that it was not a case similar to that of Guinon and Meunier[1] in which both acute tuberculosis and typhoid were found at autopsy, the typhoid being identified only by the presence of the bacillus of Eberth in the spleen and lung? In the latter case, had no cultures been taken the diagnosis would certainly have been considered miliary tuberculosis. The absence of macroscopic typhoid lesions at autopsy can no longer be considered proof of the absence of a typhoid infection. The case recently reported by Flexner and one which I had an opportunity of seeing last year with Dr. J. H. Wright of the Massachusetts General Hospital are examples of undoubted typhoid fever without any characteristic intestinal lesions. The lesson of such cases is that when a patient presents a serum reaction in his blood during life and no ulcerations are found in the intestines after death—such a case cannot be counted as a failure of the serum test. Unless the spleen, liver, mesenteric glands, and gall-bladder have been carefully examined by cultural methods for typhoid bacilli and their absence determined, the evidence of the autopsy is insufficient. In many published cases purporting to exemplify the presence of a serum reaction in diseases other than typhoid, the autopsy is either wanting altogether or lacks cultural experiments.

Further, in rare cases even cultural tests may prove negative, and yet the case may be one of typhoid fever in which the organisms have died out. I remember a case of undoubted typhoid in which post mortem only a single colony could be made to grow from the organs. Suppose this colony had not been found; there would have been no positive evidence for the diagnosis of typhoid except the serum reaction and the clinical symptoms. Yet no one would have been right in saying that the case was typhoid. This may be the explanation of v. Oordt's case (*Münch. med. Woch.*, 1897, No. 3), in which a serum reaction was present during life even with a dilution of 1:40, yet at autopsy only pneumococcus meningitis was found.

Besides the cases already analyzed, positive serum reactions in non-typhoid cases have been reported as follows:

Appel and Thornburg (*Journal of the American Medical Association*, February 6th, 1897) got clumping with blood of a case

[1] Guinon and Meunier: Soc. méd. des Hôpitaux, April 7th, 1897.

of grippe, which, however, occurred in a patient who had had typhoid eight years previously. The same observers found that rabbits' blood and hens' blood would occasionally agglutinate typhoid bacilli.

Block (*Johns Hopkins Bulletin*, November, 1896) found a marked reaction in the blood of a case of pernicious malaria (moribund) and a partial reaction in a case of diabetic coma. The dilution was 1:16 and the time limit thirty minutes. Motility was not arrested.

Biggs and Park (*American Journal of the Medical Sciences*, February, 1897) out of 187 control cases got positive reaction in a case of typhus fever. There is possibility of a mistake in the diagnosis, but Colville (*British Medical Journal*, October 16th, 1897) also found agglutinative power in the blood of a case of typhus fever.

Catrin (*La Semaine médicale*, 1896, p. 410) found a positive clump reaction in a case of malaria. The patient, however, had had typhoid five years before.

Elsberg (New York *Medical Record*, April 10th, 1897) out of 148 control cases got one positive reaction in a case of gall-stones. In view of the not infrequent association of gall-stones and cholecystitis with the presence of typhoid bacilli in the gall-bladder, it is quite possible that the positive reaction may have been no exception to the law that a serum rection means a typhoid infection somewhere.

Förster (*Zeitschrift für Hygiene*, etc., Vol. XXIV.) found 1 positive case among 90 controls.

Gehrmann and Wynkoop (Chicago *Medical Recorder*, May, 1897) reported 11 positive reactions among 56 control cases. This was in public health work, under conditions which make it difficult to be accurate in technique and to exclude a previous typhoid infection.

Grünbaum (*Lancet*, December 19th, 1896) found that he could get clumping with the blood of a parturient woman, provided he mixed it with equal parts of culture. When properly diluted there was no considerable reaction.

Johnston (New York *Medical Journal*, January 16th, 1897, and in other articles) reports that with certain cultures he can obtain clumping, even with normal blood (dried). He has *never found any clumping power in any non-typhoid blood provided attenuated cultures are used.* This fact is of the greatest importance, as Johnston's experience is probably greater than that of any other observer. A year ago (June, 1897) his individual experience covered 600 cases. By this time he must have examined over 1,000 cases.

Le Fevre (New York *Medical Journal*, March 27th, 1897) got "positive reaction" in pneumonia, cancer, and heart disease. He used suspensions instead of cultures, and one cannot help thinking that the failure satisfactorily to dissociate the bacilli may have produced the supposed clumps.

J. L. Miller (Chicago *Medical Recorder*, May, 1897) found 2.5 per cent of positive clump reactions in his 105 control cases. He used dried blood in most of these cases, and did not himself feel satisfied with the technique—no definite dilution being made. Rheumatic affections seem to Miller especially likely to show pseudo-reactions.

Park (New York *Medical Journal*, March 27th, 1897) found clumping in one case of cancer of the intestine.

Reed (*Johns Hopkins Hospital Bulletin*, March, 1897) got similar results in a case of tuberculosis. In neither of these cases could typhoid, past or present, be positively excluded.

G. B. Shattuck (*Medical News*, May 8th, 1897) reports positive reactions in the blood of two negroes neither of whom was aware that he had ever had typhoid.

Brannan (New York *Medical Journal*, March 27th, 1897) has likewise found positive reactions in four negroes who had never had typhoid so far as they were aware. The only case of a positive reaction outside of typhoid in my own experience was in a negro. Brannan's cases were cirrhotic liver, nephritis, puerperal sepsis, and diabetes; mine was pernicious anæmia. Eleven negroes tested by Brannan, eleven by G. B. Shattuck, and seventeen by myself, showed no reaction. Brannan's dilution was only about 1:3.

This apparent tendency of negroes' blood to give pseudo-reactions with typhoid bacilli has never been confirmed in any considerable number of cases, yet it has already given rise to a good deal of speculation as to the possibility that negroes are immune to typhoid and manifest their immunity in this way. The whole subject needs further investigation.

Wilson (Philadelphia *Medical Journal*, 1898, No. 13) reports positive reactions in cases diagnosed as grippe, acute mania, poliomyelitis, and doubtful reactions with cases of phthisis and grippe. In none of these cases was the history of previous typhoid obtained, although it could not be excluded in any of them.

J. B. Thomas (*Medical News*, April 23d, 1897), using dried blood and no definite dilution or time limit, got clumping with the blood of a single case each of malaria, tuberculosis, and rheumatism.

Villiès and Battle (*Presse Médicale*, October 10th, 1896) also report a positive reaction in malaria. I know no details as to their technique, and there was apparently some doubt as to diagnosis ("typho-malaria" from Madagascar).

Ziemke (*Deut. med. Woch.*, April 8th, 1897) is sceptical as to the usefulness of the reaction on account of the numerous "pseudo-reactions" found by him in other diseases (hysteria, phthisis, rheumatism, etc.).

My own results in 625 control cases were all negative except in a single instance; a negro with pernicious anæmia showed a prompt reaction with a dilution of 1:10. The patient had no remembrance of any continued febrile disease. I never tested his blood a second time nor made any experiments with higher dilutions. I think the chances are that I made some slip in technique.

Johnston in "many hundred non-typhoid cases" has never yet seen a positive reaction under conditions where typhoid past or present could be excluded.

The six per cent of positive reactions reported in cases other than typhoid are to be explained, I believe, partly by errors of technique, and partly by the existence of foci of more or less "healed" typhoid infection within the body, which react to the test as foci of "healed" tuberculosis react to tuberculin.

On the whole I am convinced that a positive reaction always means typhoid infection past or present, provided the test is properly performed. But we must never forget, as we are too apt to do, that a positive serum reaction even in dilutions of 1:800 may mean only the leavings of past typhoid infection, since Widal has seen such a reaction eight years after infection. Again, it is equally certain, in view of Cushing's case of typhoid cholecystitis with positive serum reaction but without any history of typhoid obtained (see page 106), that our notions of "*typhoid infection*" have got to be enlarged.

Not only can we have typhoid infections in the lung, the

brain, the liver, the bladder, and various other organs, but we can have typhoid without any intestinal lesions at any time, as Flexner and others have shown.

Therefore given a case of fever and a positive typhoid serum reaction, we have always to bear in mind that the typhoid infection which the reaction proves may be one smouldering in some internal organ such as the gall-bladder, the bone marrow, or the urinary bladder, and due to an attack the history of which may or may not be obtainable. Luckily this set of conditions is a very rare one. The great majority of cured typhoid patients show no reaction after a year or two, and positive clumping almost always means present typhoid. Still the possibility remains and must not be forgotten.

CHAPTER VIII.

CASES ILLUSTRATING THE VALUE OF SERUM DIAGNOSIS IN TYPHOID FEVER.

CASE I.—In July, 1898, one of the volunteers for the war with Spain was brought back to his home in Boston suffering from a disease which had been pronounced "malaria" by four different physicians who saw him at intervals during his journey north. *No one had examined his blood.* A specimen taken the day after his arrival in Boston showed a prompt and typical agglutinative reaction in a dilution of 1 : 150. Yet he had been drenched with quinine all along, to the great detriment of his strength, under the idea that he had malaria. No malarial organisms could be found in his blood despite careful search, and the course of the disease established the diagnosis of typhoid.

CASE II.—Proved by autopsy to be malignant endocarditis, which had been my diagnosis throughout on account of the negative serum reaction and the persistent marked leucocytosis. Some of the most noted clinicians in this country had pronounced the case typhoid. In several other cases of malignant endocarditis in which I have examined the blood the association of marked leucocytosis with negative serum reaction has greatly assisted the diagnosis.

CASE III.—In May, 1897, I was testing in a routine manner the blood of every case in the Massachusetts General Hospital exclusive of typhoid-fever cases, in order to see how frequently the agglutinative reaction might occur in other diseases. Among 40 cases examined one day there was one marked positive reaction and 39 negative ones. On looking up the record of the positive case I found a clinical picture as unlike typhoid as could be conceived. The patient had been in the hospital a week at the time of my test, and her temperature had been *subnormal* all that time. The symptoms were those of a mild grade of melancholia accompanied by some delusions. Vomiting was conspicuous and anorexia complete. She was very thin, weak, and pale. On account of her mental condition no previous history could be obtained. She gradually lost strength and died of malnutrition two weeks later. At the autopsy there were found healed typhoid ulcerations in the intestine, and cultures from the spleen showed colonies of bacilli responding to every test for the identification of typhoid bacilli and strongly clumped by the serum of other typhoid patients. Information secured after her death showed that she had been normal mentally until after an attack of fever, from which she was convalescent at the time she entered the hospital. The diagnosis of post-typhoidal insanity which I was enabled to make, owing to the results of the serum test, could not have been made, so far as I can see, in any other way.

CASE IV.—Not long after this I was given a specimen of blood dried on paper with the following message: "This patient has valvular heart disease. The temperature has been a little irregular of late, and her family physician, with whom I saw her in consultation, thought it might be typhoid. I don't think there is a chance of it, but I took some blood just to reassure him." On testing the blood I found an instantaneous reaction with a dilution of about 1:100. The patient afterward had an intestinal hemorrhage, and showed other unmistakable signs of typhoid fever.

CASE V. (From Bensaude).—A girl of 21, with a good family history and negative past history. Two weeks before entering

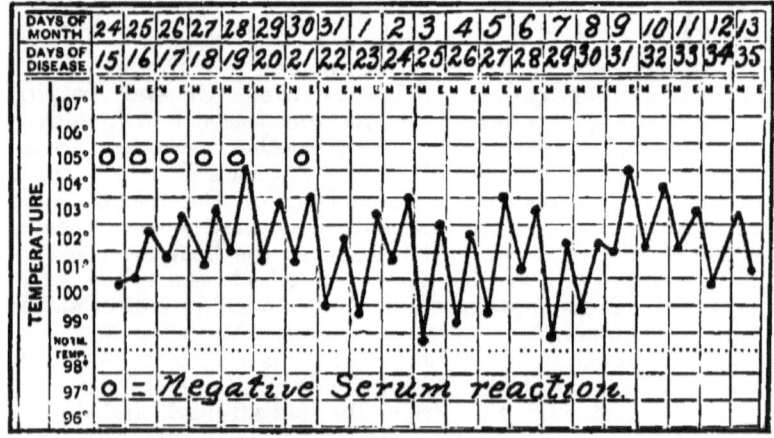

CHART VIII.—Case V.

the hospital she began to feel tired and to lose appetite. Headache, thirst, constipation, indefinite abdominal pains, vertigo and increasing apathy finally brought her to the hospital for advice, July 24th, 1896. For the last five days before entrance she had diarrhœa (five or six stools a day) and slight epistaxis. There was no cough.

At entrance the spleen was enlarged, the tongue coated, the mental condition typically "typhoidal." There were no leucocytosis and no Widal reaction. The chart during the first week of her stay in the hospital is shown above. Physical examination was negative, except as above noted.

The serum reaction was tried on the 24th, 25th, 26th, 27th, 28th, and 30th of July, and was each time entirely negative 1:10. Puncture of the spleen on the 3d of August drew only sterile blood. About this time the diarrhœa ceased and the patient began to have dry cough. The lungs were still negative. A few days before death, which occurred on the 16th of August —twenty-four days after entrance—fine râles were heard over the left apex, later extending over the whole left lung. There were no signs of consolidation, but dyspnœa became intense. Autopsy: miliary tuberculosis of lungs, liver, spleen, and kidneys.

In this case one would certainly have felt confident of the diagnosis of typhoid, but for the absence of serum reaction.

CASE VI.—At the request of Dr. Greene of the Massachusetts Eye and Ear Infirmary I examined last year the blood of a patient in that institution who, while convalescent from an operation on the mastoid process, had begun to run a temperature very strongly suggestive of typhoid fever. It was in fact one of those "typical" typhoid charts which one often sees in text-books articles on typhoid fever. The patient's operation wound was nearly healed, perfectly clean and healthy looking, and he was entirely free from pain there or elsewhere. His spleen was enlarged on percussion, and there was an abundant crop of rose spots on his abdomen. The diagnosis of typhoid had been made by several excellent clinicians, but as the patient had been in the hospital for some weeks when the fever began it could only be supposed that he had contracted the disease in the hospital itself. This seemed unlikely, as it was February and there were no other cases of typhoid in the hospital.

The examination of the blood showed no serum reaction and a marked leucocytosis. I made the diagnosis of septicæmia.

At autopsy thrombosis of the cerebral sinuses with extension into the jugular, and septic softening of the thrombus were found. No evidence whatever of typhoid.

CASE VII. (Achard and Castaigne, *Gazette Hebdomadaire*, July, 1897).—A man of 37 entered the hospital after a week's debauch complaining of headache, insomnia, vomiting, diarrhœa, and fever. He was not at all stupid in aspect, coughed much, and after a few days in the hospital signs of consolidation were detected at the base of the left lung. Puncture of this part of the lung showed no pneumococci, but typhoid bacilli were cultivated from the fluid aspirated and also from the spleen. As in many cases in which there is evidence that bacilli have been carried about by the blood and caused metastasis, the serum reaction appeared late, after the temperature reached normal.

CASE VIII. (Bruhl: *Gazette Hebdomadaire*, January 31st, 1897).—A man of 32 entered presenting every appearance of pneumonia. His illness began abruptly four days earlier with a chill, dyspnœa, and pain in the left side. His sputum was rusty and viscous. But two days after entrance the serum reaction was positive, and three days later still the typhoid bacillus was obtained in pure culture in the blood withdrawn from the lung, while no pneumococci were discovered. Some of the sputum injected into a mouse caused death in sixty hours, but the blood and organs of the mouse showed only Eberth's bacilli and no pneumococci. Later the patient had a relapse and an intestinal hemorrhage.

Wilson relates the details of a case of supposed appendicitis in which the presence of the clump reaction did not prevent the surgeon from operating. A normal appendix was found and the case turned out to be typhoid fever.

Achard, Catrin, Lemoine, Villiès and Battle, and others have borne testimony to the value of the reaction in differentiating typhoid from attacks of gastro-enteritis with fever.

Haedke mentions the following interesting cases:

CASE I.—A girl who had had attacks of "colic" every three weeks for years entered the hospital with an attack much like the rest, accompanied by jaundice, vomiting, and soreness in the epigastrium. Temperature, 105.2°. Serum reaction positive. Autopsy: typhoid fever.

CASE II.—Woman three weeks post partum. Chills, vomiting, intermittent uterine pain. Serum reaction positive. Turned out to be typhoid.

CASE III.—After a few days of depression a man attempted suicide by drowning. After being brought to the hospital he had fever for a few days and then the temperature became normal. Serum test positive. Relapse and typical typhoid course.

The following cases which I was able to follow in the wards of the Massachusetts General Hospital illustrate some of the forms of help to be derived for the serum reaction (see also next chapter.

CASE I.—Mary Connell, 52, wife. Family history good. Cough and irregular chills for six months. At entrance was fairly nourished. No cough to speak of. Breathing over *right* apex harsh and voice sounds increased. Edge of liver felt below ribs. Examination otherwise negative. No diazo. Leucocytes,

VALUE OF SERUM DIAGNOSIS IN TYPHOID FEVER. 93

3,500. No malarial organisms. Later the spleen became enlarged and diarrhœa set in. *Serum reaction negative.* Later, bacilli were found in the sputa.

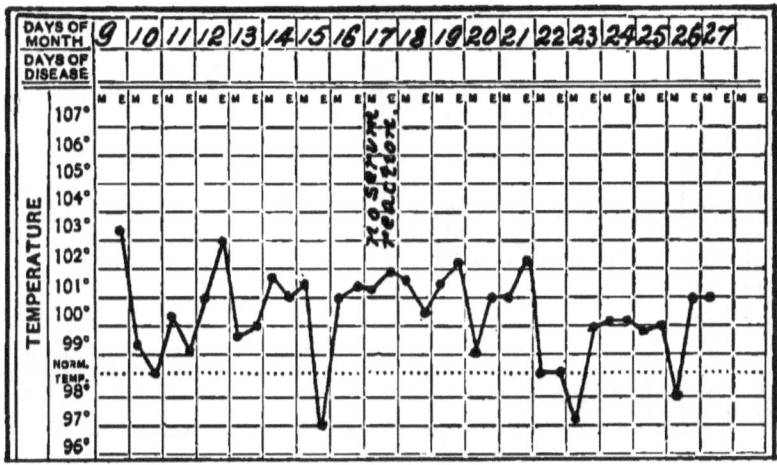

CHART IX.—Case I.

CASE II. *Abortive Typhoid.*—Jackson, 18, porter, two and a half weeks' malaise, one week in bed. At entrance spleen was

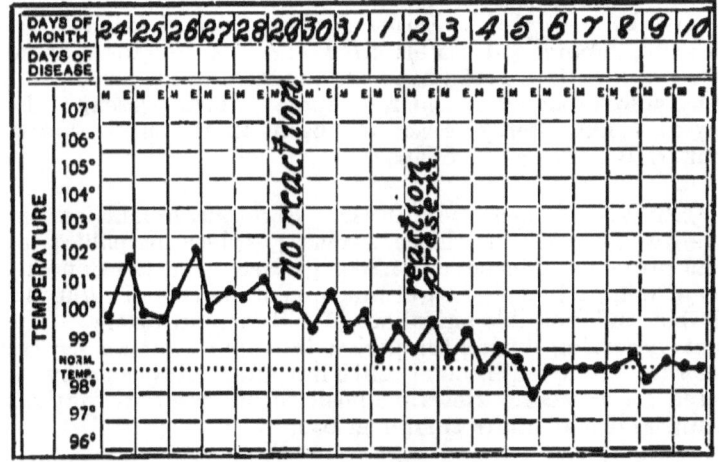

CHART X.—Case II.

enlarged, no rose spots, no diazo; leucocytes, 5,200. Reaction absent five days after entrance, present ten days after entrance.

CASE III. *Abortive Typhoid.*—Miss Otis, 30, nurse. Malaise four days, bed one day. No splenic tumor, rose spots,

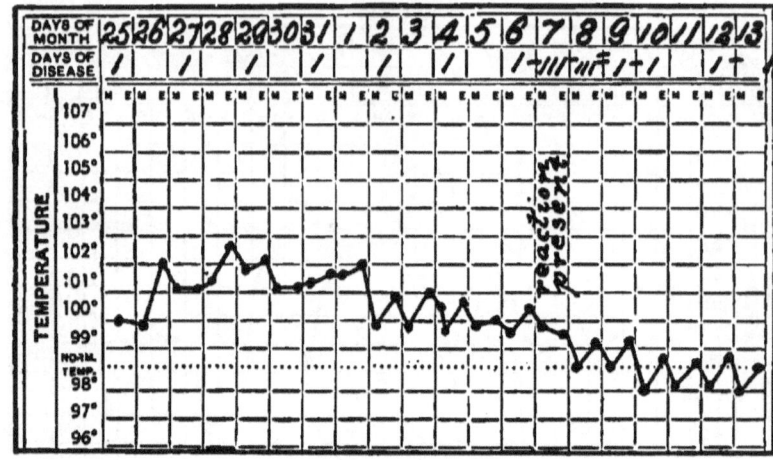

CHART XI.—Case III.

nor diazo at any time. Leucocytes, 3,700. Serum reaction present on thirteenth day after entrance. Later had a relapse.

CASE IV. *Typhoid; Serum Reaction Absent till Relapse.*—Girlando, laborer, 21. Felt tired and "mean" for a month. Gave up work two weeks ago; was nine days in bed. At entrance spleen palpable; no rose spots; diazo present. Leucocytes, 4,500. Serum reaction absent in the original attack (four trials), present in relapse.

CASE V.—Holmes, 34, driver. Eight days' malaise; four days in bed with headache, nausea; spleen palpable; rose spots present. Leucocytes, 4,200. Diazo present. Consolidation of left, and later of right lung. Thrombosis of right femoral vein and possible embolism of left femoral artery. These events occurring during the period in which I got *no serum reaction* (on sixth, tenth, and twelfth day after entrance) may have caused its absence. Reaction positive on twentieth day after disappearance of above complications.

CASE VI. *Perinephritic Abscess Simulating Typhoid.*—Jeffries, 39, laborer. Four weeks' cramps in belly, but worked until one week ago when he went to bed with diarrhœa and weakness, which have increased since. No cough. Physical examination negative except for slight enlargement of liver. No diazo; no tenderness or any localizing symptom or sign. Leucocytes, 15,300. No serum reaction on sixth day after entrance nor on ninth. No chills or sweating. Later began to spit pus, and a perinephritic abscess was found and opened; recovery.

CASE VII. *Fatal Typhoid; No Serum Reaction.*—Hoyt, 19, student. Ten days malaise; four days in bed on account of headache. At entrance no rose spots or splenic tumor; diazo present. Leucocytes, 6,500. Three days later rose spots appeared. On this day the *serum reaction* was *negative*. Two days later perforation, operation, death. Autopsy: Bacilli of Eberth in spleen showed hardly any motility and soon died out on agar-agar despite of weekly transplantations.

CASE VIII. *Retrospective Diagnosis of Typhoid by Serum Reaction.*—Adam Hay, 33, typesetter. Poorly for months. Ten days in bed. At entrance abdomen was distended. No rose spots or splenic tumor; no diazo. Leucocytes, 4,500. Serum reaction two months after entrance, during second consecutive relapse—forty-eight days after entrance.

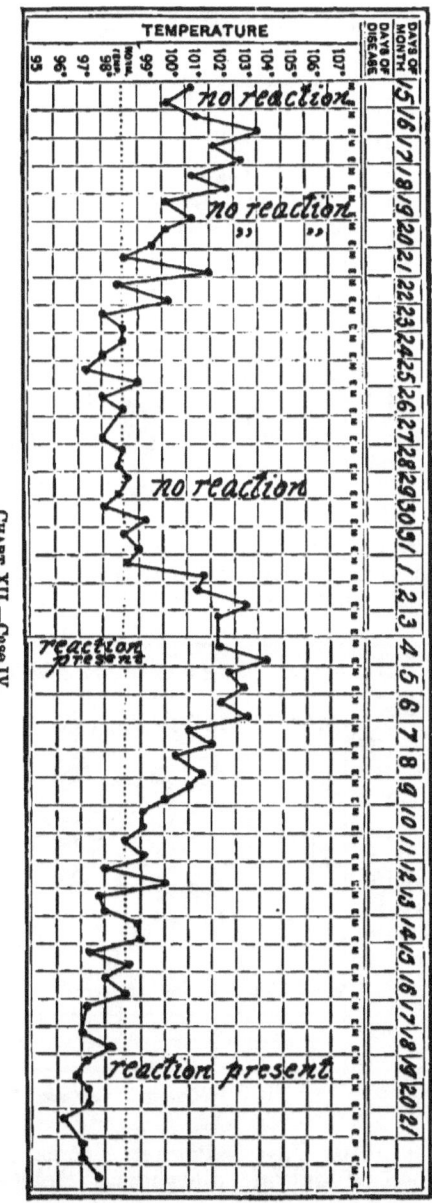

CHART XII.—Case IV.

96   THE SERUM DIAGNOSIS OF DISEASE.

CASE IX.—Leveton, 21, female. Four days' malaise; not in bed. No spleen or spots at entrance. Urine negative; diazo present. Leucocytes, 5,050. *Serum reaction was obtained*

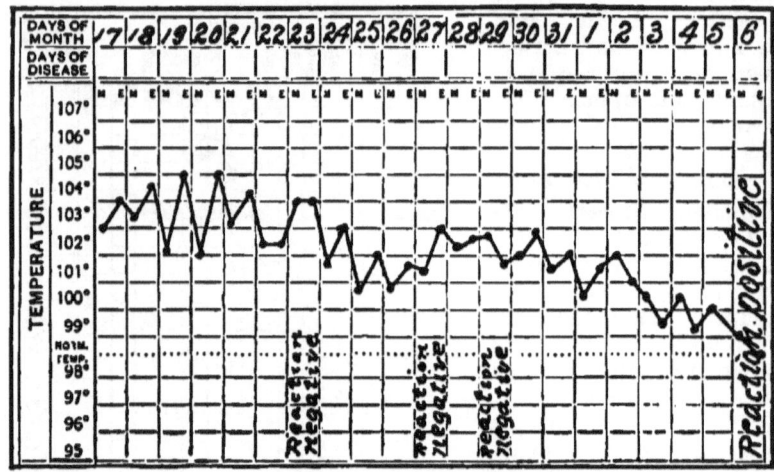

CHART XIII.—Case V., page 94.

at once, and *two days later* the splenic tumor and rose spots appeared. Ran a normal course with a relapse. Discharged on forty-third day.

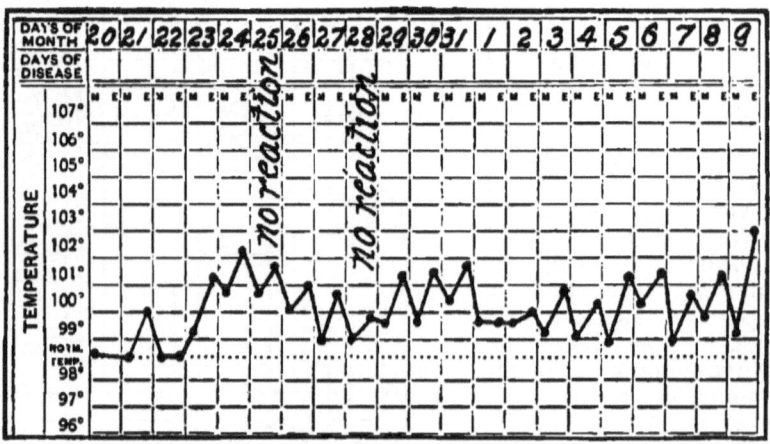

CHART XIV.—Case VI., pages 94-95.

CASE X.—Gallagher, fireman, 27. Three weeks' fever; in bed off and on; diarrhœa. Spleen palpable at entrance and spots present. Urine negative; no diazo. Leucocytes, 4,300. *First serum test* fourteen days after entrance, when he had been afebrile three days. *No reaction.* *Second serum test* during relapse showed on twenty-eighth day after entrance *a very prompt reaction;* diazo present only in relapse. Discharged on forty-fourth day.

CASE XI.—Doherty, mason, 29. One week's malaise, etc.; not in bed. At entrance, spleen palpable, rose spots present, typhoidal expression; no diazo. Urine negative. Leucocytes, 7,200. *Serum reaction on second day after entrance;* rather slow in appearing, but typical. Second test on seventh day after entrance, positive. A very mild case.

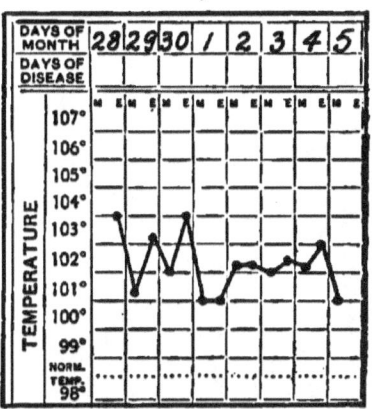

CHART XV.—Case VII., page 95.

CASE XII.—Kewick, 21, glazier. Four days' malaise, one day diarrhœa. At entrance no spleen or spots; no diazo;

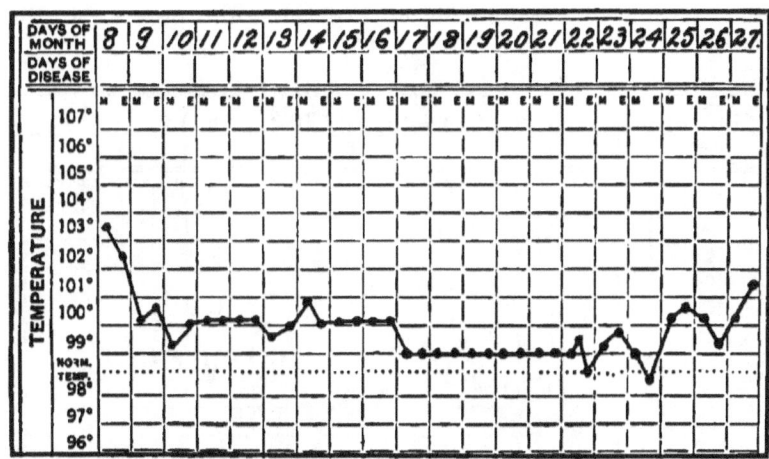

CHART XVI.—Case VIII., page 95.

headache marked. Leucocytes, 5,400. Seventh day after entrance rose spots appeared. Ninth day after entrance serum test positive. Diazo appeared two days later.

CASE XIII.—McIsaac, 34, laborer. Malaise one week. Diarrhœa three days; not in bed. At entrance spleen felt; no diazo. Leucocytes, 3,900. No more diarrhœa. Serum test

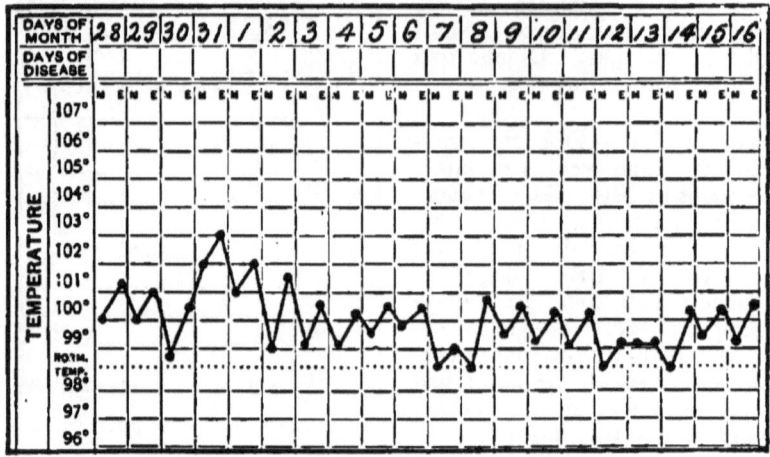

CHART XVII.—Case VIII. (continued).

positive on second day after entrance and again on tenth day after entrance. Discharged on twenty-eighth day after entrance.

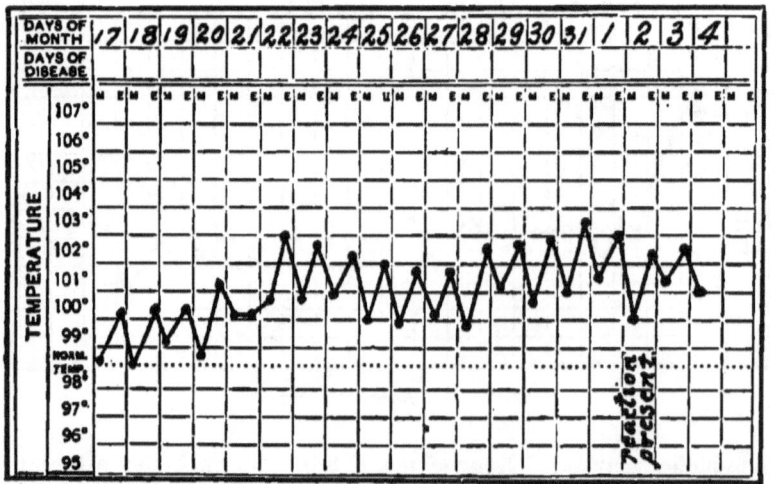

CHART XVIII.—Case VIII. (continued).

CASE XIV.—McDonald, 24, lineman. Previous family history negative. Five days' malaise; in bed two days. Head-

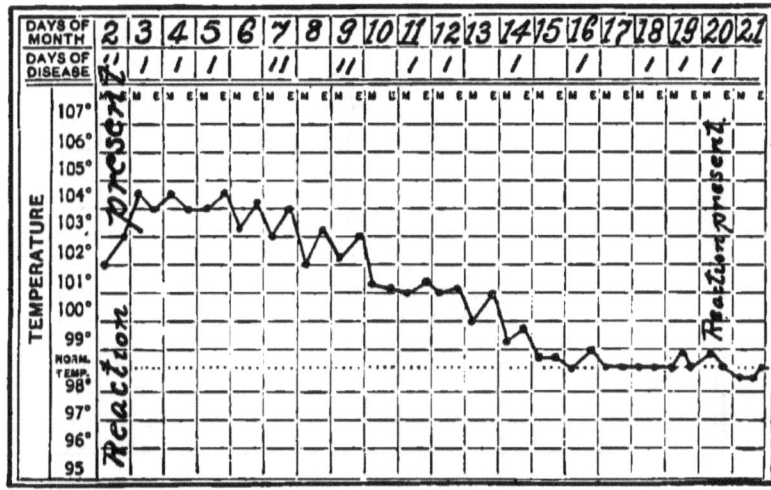

CHART XIX.—Case IX., page 96.

ache, stiff neck, and muscular soreness, nausea. Now (at entrance) feels brighter and has more appetite; bowels regular;

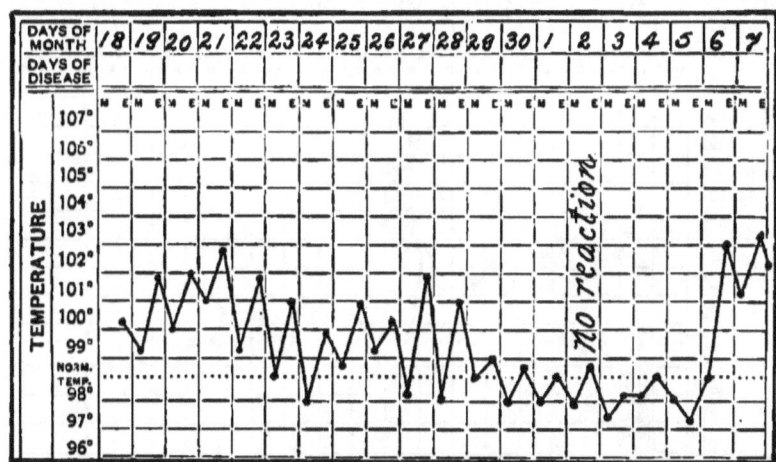

CHART XX.—Case X., page 97.

no splenic enlargement; no rose spots. Leucocytes, 5,500. No diazo; serum reaction present five days after entrance. *Diagnosis: Abortive Typhoid.*

CASE XV.—McConlogue, 23, waitress. Two weeks' malaise; five days in bed with fever, chills, diarrhœa, cough,

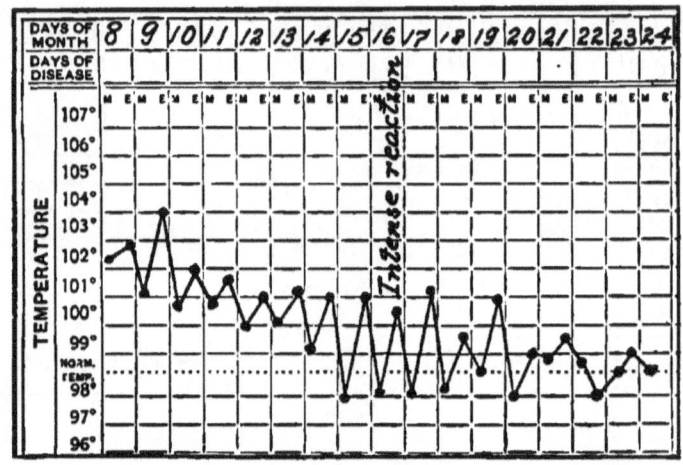

CHART XXI.—Case X. (continued).

nausea, night sweats. At entrance no splenic enlargement nor rose spots, and none were ever found; diazo present. Leu-

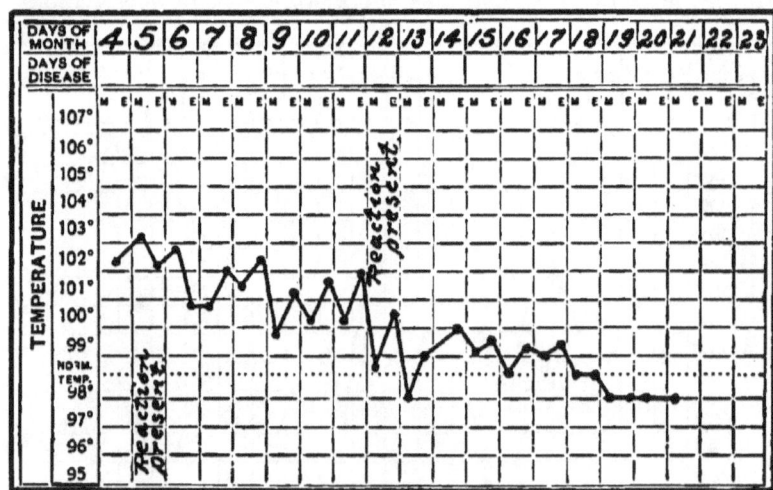

CHART XXII.—Case XI., page 97.

cocytes, 2,200. Nervous symptoms very marked, subsultus delirium, œdema of lungs, death. No autopsy. Serum reaction not found.

CASE XVI.—Delano, 26, clerk. Ten days' malaise, six days in bed. At entrance mentally dull and confused. Splenic area

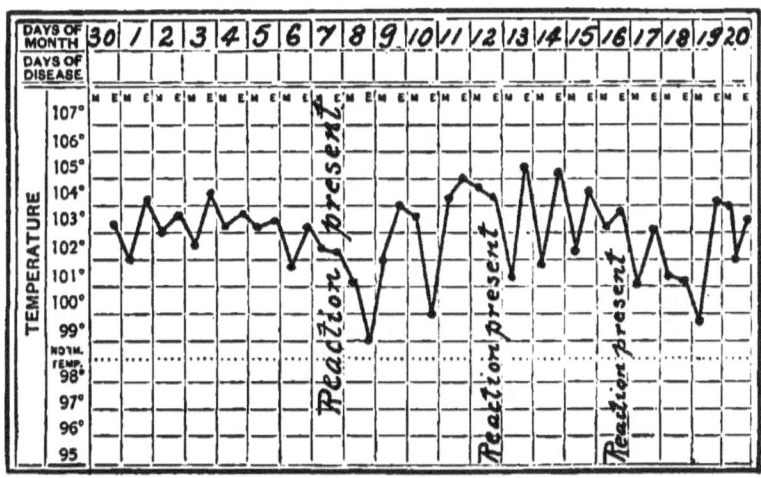

CHART XXIII.—Case XII., page 97.

tympanitic; rose spots abundant; diazo present. Leucocytes, 2,400. Serum reaction on fourth day after entrance, but not on third.

CASE XVII.—Erichson, 27, wife. Nine days' chills and malaise, three days in bed. Spleen felt; no rose spots; diazo

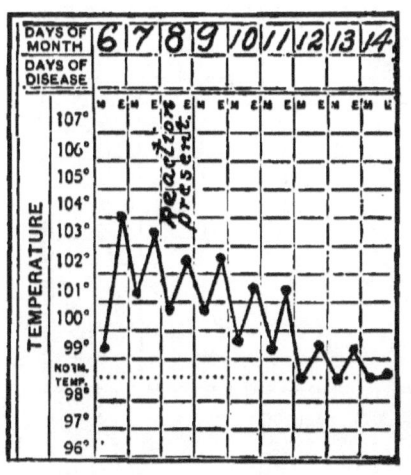

CHART XXIV.—Case XIII., page 98.

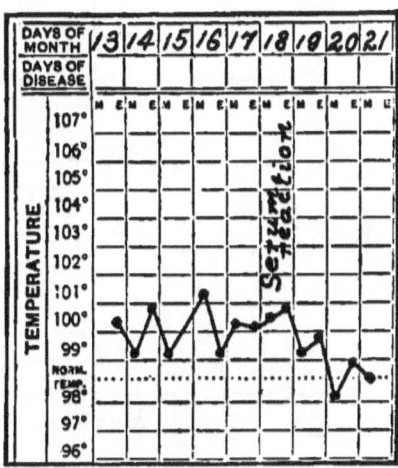

CHART XXV.—Case XIV., page 99.

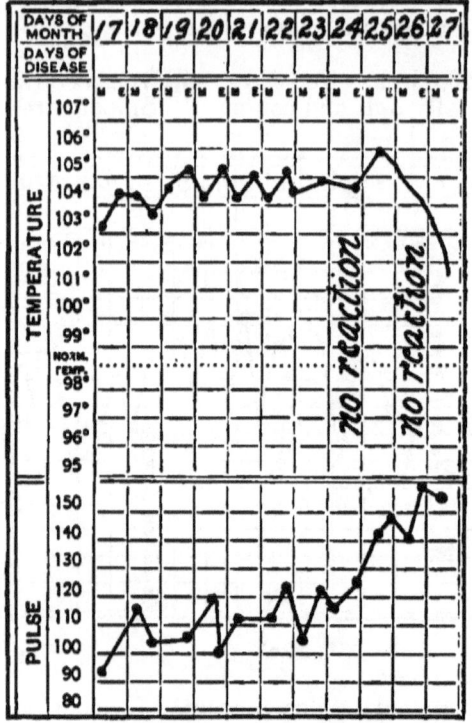

Chart XXVI.—Case XV., page 100.

present. Leucocytes, 7,600. Ran a typical course. Serum reaction absent on the eighteenth day after entrance; present on nineteenth and twenty-second days.

Case XVIII.—McClinchey, 21, wife. Three weeks' malaise; in bed half a day. No splenic enlargement or spots; diazo present. Leucocytes, 4,000. *Reaction present at entrance.* Worked till day before entrance.

Case XIX.—Crimmins, 30, waiter; five to six whiskeys a day. Five weeks ago noticed pain and swelling of abdomen. Nausea and diarrhœa for two weeks; in bed three weeks. Now com-

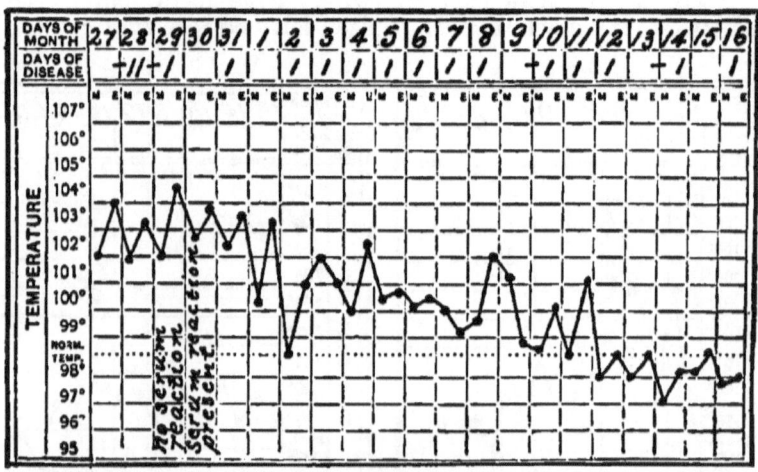

Chart XXVII.—Case XVI., page 101.

plains of dyspnœa, anorexia, palpitation. Lost from ten to fifteen pounds. At entrance pale and pasty, stupid. Tongue tremulous, dry, and coated; sordes on gums. Liver from upper

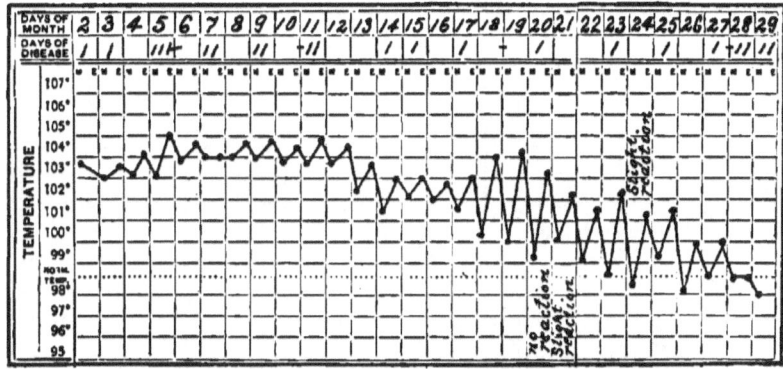

CHART XXVIII.—Case XVII., page 101.

edge of fifth rib to a point one and a half inches below rib margin. Fluid in belly. Spleen enlarged. No rose spots; no diazo.

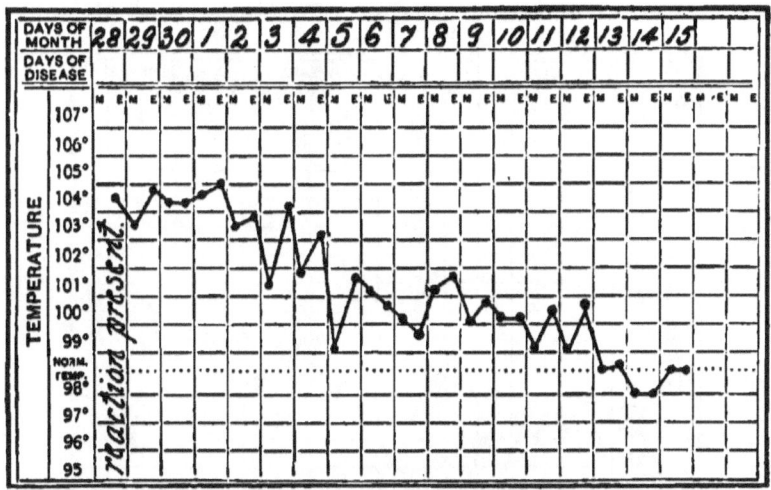

CHART XXIX.—Case XVIII., page 102.

Red cells, 2,656,000; white cells, 6,300. Diagnosis between febrile cirrhosis and typhoid. Serum reaction positive on day of entrance. Course proved the case to be typhoid.

Case XX.—Dawson, 23, female. Entered insane and without history. Physical examination negative; no diazo. White

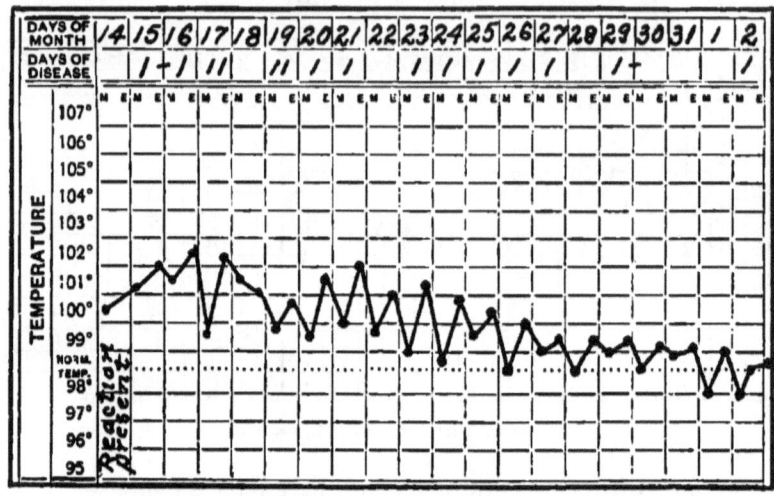

Chart XXX.—Case XIX., page 102

cells, 6,000. Serum reaction positive twice. One week later friends seen and a history obtained of a febrile attack six weeks

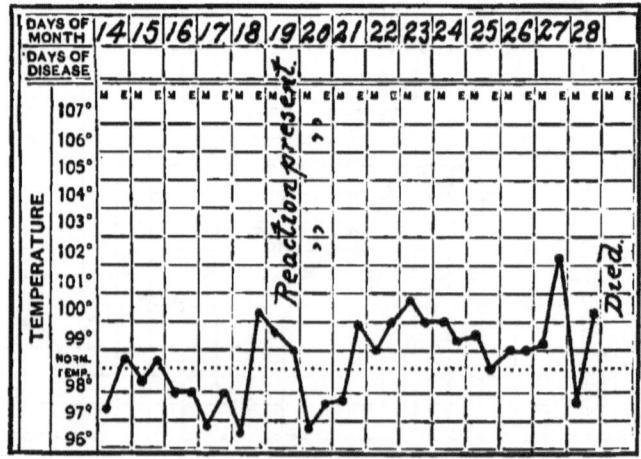

Chart XXXI.—Case XX.

before entrance. Previously sane. Fifteen days after entrance she died. Autopsy showed typhoid ulcers healing and Eberth's bacilli in spleen.

CHAPTER IX.

SERUM DIAGNOSIS IN ATYPICAL FORMS OF TYPHOID AND IN ALLIED INFECTIONS.

Retrospective Diagnosis.

"Retrospective diagnosis" (Widal), such as was exemplified in the first case above reported in which a melancholia was shown to be of the post-febrile type, may be of considerable importance in a good many cases. Thus Courmont reports a case of multiple neuritis following an attack of what was supposed to be dysentery. The marked typhoid serum reaction showed it to be a post-febrile neuritis.

Achard was able by the same test to determine the source of a focus of osteomyelitis which followed a febrile attack a year before.

Blanquinque[1] satisfied himself by serum tests made in convalescence that two cases which had been supposed to be psittacosis had in fact been typhoid (see Psittacosis, page 140). The importance of recognizing cases seen first in convalescence is evident from the point of view of public health.

Investigation of Doubtful Epidemics.

From the standpoint of public health the use of the serum reaction in convalescent or cured cases will certainly be of importance in searching out the source of the infection as well as in determining the nature of the epidemic. Fraenkel, investigating an epidemic (*Deut. med. Woch.*, January 14th, 1897) found that the two persons who were suspected of having started it a year before both showed marked positive reactions.

Smith, of the Army Medical School at Netley, England, who examined many cases at Maidstone with reference to the origin of the epidemic, noted that the serum reaction often enabled us to exculpate persons who on account of a diarrhœa

---

[1] Blanquinque: Gazette Hebdomadaire, February 4th, 1897.

might have been supposed to have had a typhoid and blamed for spreading the infection. He examined the blood in such cases with negative results, and was thereby enabled to forsake a false clew for a truer one. Similarly Boyce (*Lancet*, February 13th, 1897) has investigated an epidemic of fever with dysenteric discharges containing large numbers of a bacillus closely allied to the typhoid organism. He found the blood of the cases in this epidemic had no agglutinating effect on Eberth's bacillus.

Typhoid Fever in Infancy and Old Age.

Among Wilson and Westbrook's series of nearly 700 cases of typhoid with positive serum reaction there were 32 in children under seven years of age distributed as follows: 9 cases at six years, 11 at five years, 4 at four years, 3 at three years, 4 at two years; 1 at two days.

In the observers' opinion the test appears earlier in children than in adults. Many of their earliest reactions were in infants. This is of great importance, since the diagnosis of typhoid in infancy has always been a stumbling-block to clinicians.

Widal, Haushalter, and Conture have reported cases in which the diagnosis of typhoid in infants was made possible by the serum test. Achard has found it equally helpful in old age.

The Serum Reaction in Abortive Cases.

H. M. Bracken (Philadelphia *Medical Journal*, Vol. I., No. 8, 1898) describes a case in which the illness lasted but a few days, yet the Widal reaction was present on the fourth, tenth, seventeenth, twenty-sixth, twenty-eighth, sixty-fifth, and eighty-seventh days—at a dilution of $1:25$. It was sometimes present with a dilution of $1:500$.

Among the cases observed by me at the Massachusetts Hospital and described in detail above (see page 93) there are several of this type, and others of a similar nature have been mentioned by various observers.

Typhoid Infection without Fever.

Bondet (*Soc. Nat. des Sciences de Lyon*, February 15th, 1897) describes a most unusual case in which the diagnosis would have been impossible but for the serum reaction. The patient was a woman who had had charge of three children ill with typhoid. Her complaints were of headache and backache *without any fever*, without diarrhœa, rose spots, enlarged spleen, or any of the ordinary signs and symptoms of typhoid. Her occupation as nurse of patients ill with typhoid suggested that the serum test should be tried. It was prompt and positive on three occasions. Yet the apyretic condition continued, and after six days of normal temperature the diagnosis of typhoid was considered so unlikely that the patient was given solid food. That evening she had a perforation of the intestine, and on autopsy typical typhoid ulcerations were found.

Bensaude likewise had two cases of typhoid in which there was no fever at the time of the test (positive), though later the temperature rose.

Courmont mentions a case (*Thèse de Lyon*, July 9th, 1897) in which the temperature oscillated throughout the whole illness between 98.6° and 103.1° without any true continued fever. The diagnosis was established through the serum reaction.

In another case described in this thesis there was a hyperpyrexia so great that the diagnosis of typhoid would have been doubtful but for the serum reaction (positive).

The Serum Reaction in Cases of Typhoid—without Intestinal Lesions.

Twelve cases are on record. In four the serum reaction was tested and found present. In addition to these I recollect a case at the Massachusetts Hospital in 1897 in which Dr. J. H. Wright could not recognize anything distinctive of typhoid in the slight intestinal lesions present at autopsy, but in which cultures from the spleen were decisive.

In Pick's case (*Wien. klin. Woch.*, 1897, p. 82) the only symptoms were fever, bronchitis, and constipation. There was no enlargement of the spleen and no rose spots. The Widal reaction was marked, and the examination of the stools by Elsner's method showed the presence of typhoid bacilli, but at autopsy no lesions were found in the intestine. Cultures from the spleen showed the presence of typhoid bacilli.

In Guinon and Meunier's case (*Soc. méd. des Hôp.*, April 2d, 1897) there was a double infection with tuberculosis and typhoid, but without any characteristic intestinal lesions.

The Serum Reaction in Cases of Typhoid Masked by Another Infection.

Besides Guinon and Meunier's case already referred to, there are on record a few other cases of double infection in which the serum reaction aided the diagnosis. Chantemesse and Ramond (*Soc. méd. des Hôp.*, June 18th, 1897) describe a case with all the symptoms and signs of tuberculous meningitis. The serum reaction four days before death was positive and marked. At autopsy both meningeal tuberculosis and typhoid lesions were found.

Ziemcke has likewise recorded a positive serum reaction in a case of typhoid complicated with diplococcus pneumoniæ verified at autopsy.

Serum Diagnosis of Cholecystitis and Cholelithiasis.

The frequent association of cholecystitis and cholelithiasis with typhoid bacilli in the gall-bladder has attracted much attention of late. Flexner found typhoid bacilli in the gall-bladder in fifty per cent of the autopsies on typhoid, and Cushing[1] has recently collected six cases of post-typhoidal cholelithiasis with typhoid bacilli demonstrated in the gall-bladder in pure culture. In three of these in which the patient's blood was tested a positive serum reaction was found—in one case (v. Dungern's) fourteen years after the attack of typhoid. In Cushing's case *no history of typhoid could be obtained* in spite of most careful questioning; but the bacilli were proved to be typhoid bacilli by every known test, and the blood serum was positive at 1 : 30.

The hypothesis offered by Richardson[2] to explain the association of gall-stones with typhoid bacilli in the gall-bladder is that the bacilli are clumped in the gall-bladder by the agglutinative power of the bile, and the clumps, getting larger and larger, finally form the nucleus of a gall-stone. In Cushing's and Richardson's cases the clumping of the bacilli in the gall-bladder was demonstrated.

In view of these discoveries it seems important that in future

[1] Cushing: Johns Hopkins Hospital Bulletin, May, 1898.
[2] Richardson: Boston Medical and Surgical Journal, December 16th, 1897.

cases of suspected gall-stones should be tested for the agglutinative serum reaction.

A further source of interest in these discoveries is the fact that several of the cases reported as positive reactions in disease other than typhoid have been in gall-bladder affections.

For example, among Elsberg's 148 non-typhoid cases there was one (*a case of gall-stones*) which reacted positively over and over again, although no history of a post-typhoid fever could be elicited. In the light of Cushing's observations it seems more than likely that Elsberg's case was, like Cushing's, a primary typhoid infection of the biliary tract with resulting cholelithiasis. The cases of "jaundice" reported by Grünbaum as giving a positive reaction may possibly be explained similarly.

Serum Reaction in Meningitis Due to Eberth's Bacilli.

Troisier and Sicard report a case (*Soc. méd. des Hôp.*, January 21st, 1897) which when first seen had no fever but showed evidence of nephritis, and also a retraction of the head, with headache. The serum test was positive.

On autopsy there were found meningitis due to the typhoid bacilli, an abscess of the kidney containing a pure culture of typhoid bacilli, and healed typhoid ulcerations in the intestine.

[In the discussion of this case Ferrier mentions another case seen by him in which fever was absent, but in which the positive results of the serum test were borne out by the later course of the disease.]

Serum Reaction in the Melancholic Forms of Typhoid.

In a recent discussion reported in the *Lyon médicale* for November 7th, 1897, Taty mentioned a case which he thought would undoubtedly have passed for ordinary insanity but for the evidence given by the serum test. He thought the serum should be tested in all acute cases of delirium, stupor, confusion, or melancholia. Pierret, in discussing the paper, stated his conviction that there were now many cases of typhoid confined in insane asylums. The case of post-typhoidal insanity which I have reported above is interesting in this connection.

The Serum Reaction as a Means of Distinguishing Typhoid from Allied Infections.

Considerable discussion has taken place regarding a series of seventeen cases recently reported by Brill (New York *Medical Journal*, 1898, No. 3), in which the course of the temperature and the general aspect of the disease were those of mild typhoid, but in which no bacilli could be obtained from the spleen by puncture nor from the stools by culture, and in which the serum test was always negative. But for these bacteriological and microscopic tests the cases would undoubtedly have been considered mild typhoid, and many have so considered them despite these findings. In the absence of autopsy data we cannot consider the question settled, but I think that in such cases the serum test is doing us a service; it tends to make us, at any rate, more wary in classing all indefinite febrile affections as typhoid in the absence of any positive data. There is no doubt in my mind that the diagnosis of typhoid has been used in many instances as a cloak for our ignorance and a handy name for a variety of separate but unknown infections. Since the influenza came to visit us many cases are called "grippe" which used to pass as "abortive typhoid" or as "febricula."

I am in hopes that the serum diagnosis, applied as Brill has done in connection with examination of stools, splenic juice, and urine, may enlighten us further along these lines.

Service similar to that of Brill has been rendered us by Durham's interesting and careful researches "On the Serum Diagnosis of Typhoid, with especial Reference to the Bacillus of Gärtner and its Allies" (*Lancet*, January 15th, 1898). Durham is of the opinion that the *bacillus enteritidis* of Gaertner has often been mistaken for the typhoid bacillus, from which it differs "only by its ability to produce gas bubbles in the presence of glucose or of muscle sugar, in possessing a greater power of overcoming the preliminary acid formed in the presence of glucose or lactose, and in reducing power."

Durham himself had in April, 1896, "an attack of ill-defined nature which I thought was typhoid at the time," after which his serum reacted well on typhoid bacilli (1:200). At that time his serum was not tested on Gärtner's bacillus, but now it has ceased to react with typhoid and yet "still reacts at 1:200 with Gärtner."

The serum of two patients from the epidemic of typhoid which raged at King's Lynn in 1897 was found by Durham to react *better* with Gärtner's bacillus than with the typhoid bacillus. Thus in one case Gärtner's bacillus was clumped in a dilution of 1:500, while with the typhoid bacillus the serum reacted no higher than 1:100. In the other case the serum reacted with Gärtner at 1:100 and failed to react with typhoid at that dilution.

Various other cases from the Maidstone or King's Lynn epidemic reacted nearly as powerfully on Gärtner bacillus as on the typhoid bacillus — *e.g.*: with Gärtner 1:100, with typhoid 1:200, Case I., thirty-eighth day; severe. With Gärtner 1:500, with typhoid 1:1,000, Case II., forty-fifth day; severe; relapse.

Durham cautiously refrains from drawing any conclusions from these figures, but they certainly suggest that what we commonly call typhoid fever may be any one of several infections or a mixture of them.

"*Paracolon Bacillus*" *Infections.*

Gwyn (*Johns Hopkins Bulletin*, 1898, No. 84, Vol. IX.) reports an interesting case which resembled typhoid clinically, but in which the serum test was negative with Eberth's bacilli. From this patient's blood a "paracolon bacillus" was isolated with which the patient's serum reacted in dilutions of from 1:150 to 1:250, and the reaction persisted in the blood for two months. All this time the serum would not clump typhoid bacilli if diluted over 1:5.

Two races of colon bacilli reacted at 1:60 and 1:100 with the patient's serum (some normal sera clumped these bacilli at 1:120). On the "paracolon bacilli" a typhoid serum from another case having an agglutinating power of 1:900 reacted as high as 1:30, but no higher. Normal serum had no considerable effect on them.

Widal and Noblecourt (*La Semaine médicale*, August 4th, 1897) report a very similar case in which the blood of a patient with symptoms suggestive of typhoid would not react beyond 1:20 with typhoid bacilli, while on "paracolon bacilli" they gave positive reaction as high as 1:12,000.

Serum Tests in "Mediterranean Fever."

Thirty-four cases were tested by Aldridge (*Lancet*, May 21st, 1898). Thirty of these gave immediate reactions with Bruce's micrococcus melitensis—the organism of Malta fever. The other four cases did not react with Bruce's coccus, but did clump with the bacillus of Eberth. In most cases the reaction appeared by the fifth day. As a rule it could be obtained at a dilution of 1:100, but slowly in the course of half an hour.

Williès and Battle (*Presse médicale*, 1896, No. 84) examined three cases of Madagascar "typhomalaria," and found negative reactions with typhoid bacilli in two and positive in one.

Serum Tests in "Mountain Fever."

C. E. Woodruff has recently investigated an epidemic occurring at Fort Custer, Montana, similar to those which have been hitherto pronounced "mountain fever." It has been supposed that Montana is wholly free from typhoid—cases passing under the names of "malaria," "Red River fever," "slow fever," or "mountain fever." In this epidemic of thirty-five cases Widal's test was positive in all but one of the twenty-eight cases in which it was tried. In the milder cases it was late in appearing. In one case it was not found till convalescence. The identity of these cases with typhoid cannot be doubted. It is greatly to be hoped, as Dr. Woodruff suggests, that other Western fevers will be investigated in this way. The ignorance of their nature has led to carelessness as regards excreta and so to unnecessary spreading of infection.

It will be much easier in the future to decide as to the true nature of the supposed cases of afebrile typhoid such as have occasionally been reported in the past.

Serum Tests in the Natives of India.

Freyer (*British Medical Journal*, August 7th, 1897) asserted that the blood of all native Indians possesses agglutinating power in health, and argued that this accounted for the absence of typhoid fever among them.

W. C. Brown (*British Medical Journal*, March 12th, 1898) contradicts this *in toto*. He made thirteen tests in healthy natives with negative results; two who were sick with symptoms resembling typhoid showed marked reactions.

CHAPTER X.

SERO-PROGNOSIS.

WIDAL has from the first steadily maintained that no indications of prognostic value could be obtained from the agglutinative reaction. He has admitted that in a rough and general way the reaction tends to be most marked in the worst cases, and at the fastigium of the individual case. But to this rule there are so many exceptions that some writers have thought that rules for prognosis could be drawn on exactly opposite lines,—the worse the case the less the reaction. This view is supported by the exploded but still current belief that the reaction expresses immunity and by certain isolated observations, but it is certainly not correct.

Widal thought that it might be stated as a tolerably sure rule that in the mild cases the reaction ceases sooner in convalescence, but this is of no use in prognosis.

Courmont (*La Semaine médicale*, 1896, p. 294) was among the first to assert his belief in the possibility of a serum prognosis. In a later paper (*Lyon médicale*, August 8th, 1897) he laid down the following rules.

1. If the clumping power rises as the temperature falls, and then falls again during convalescence, repeating the temperature curve but at a later date, the prognosis is good.

2. If the case is serious clinically and the reaction is feeble, the prognosis is grave.

3. If the clumping power is feeble at the height of the fever, the outlook is bad.

Catrin (*La Semaine médicale*, 1896, p. 418) stated that if the reaction was intense during the earlier days of the disease the prognosis was bad. Widal's experience with a case which clumped in a dilution of 1:1,000 on the sixth day in bed, yet turned out mild, certainly seems to contradict this.

Tschistovitch (*Bolnitcharaya Gazeta Botkina*, 1897, No. 51) states that the more severe cases give a mild reaction until near the close of the attack. When the reaction is well marked he thinks the disease is apt to run a favorable course. In experimental infections he found that the reaction did not appear until the leucocytes had reached their lowest point and began to rise again.

On the whole he does not attribute great prognostic value to the reaction, and in this opinion he is supported by the great majority of competent observers, including Johnston, Breuer, Biggs and Park, Block, Fraenkel, Villiès and Battle, and many others. A few, like Rochemont and Sabrazés, have thought that the reaction was simply most marked in the worst cases, and in a very general way, as already said, this seems to be true; but to this rule, as to every rule for prognosis yet made, there are countless exceptions, and Fraenkel and Breuer do not find it true at all.

Prognostic Significance of Delayed Reactions.

One point of possible prognostic significance in the serum reaction mentioned by Elsberg has not, it seems to me, received as much attention as it deserves. Elsberg noticed that in four cases in which he failed to find the reaction during the first two weeks of the disease, relapse followed. He further noted that in these cases the reaction not only appeared late in the original attack but disappeared early, lasting only from twelve to sixteen days in all.

In one of his cases reaction did not appear *even in the relapse*. A second relapse occurred, during which the agglutinating power appeared in the blood; after that no further relapse occurred.

This same phenomenon—viz., the occurrence of relapse in cases in which no clump reaction is found in the original attack—has also been noted by Widal, Eshner, Breuer, Wilson, Cahill, and Thoinot and Cavasse—each in a single case—by Biggs and Park in two cases, and by myself in three cases. This makes a total of fifteen cases in which the phenomenon has been observed, and in connection with these it seems to me reasonable to refer to an observation reported by Dunlop (*British Medical Journal*, December 11th, 1897) where the case was pronounced typhoid by three consultants who saw it at different periods in its course, yet the serum reaction was persistently absent. Relapse after relapse followed, still without serum reaction, until five relapses had occurred.

It seems to me that, although the cases here collected are too few to justify any conclusions, they are of the greatest interest and importance with reference to the possibility of a serum prognosis of relapse. Further observations on this point are greatly to be desired.

I am, however, by no means unmindful of Widal's observations, which show that high clumping power at the end of the original attack by no means insures the patient against relapse. Certainly the rule does not work both ways, but the important question is, Does it work one way? In the blood of pneumonia patients a high leucocyte count is no proof that death will not follow, yet it has been abundantly proven that a low leucocyte count does mean death in the great majority of severe cases. Here is a rule that works one way only. It may be the same with the serum prognosis of relapse from absent serum reaction in the original attack.

An explanation for the occasional absence of the serum reaction in cases of undoubted typhoid fever has been suggested to Courmont, Ménetrier, and others by cases of pleuritic effusion complicating typhoid where the fluid in the chest contained Eberth's bacilli in pure culture and no serum reaction could be obtained with it. Courmont likewise observed that the presence of typhoid bacilli in a potent typhoid serum *in vitro* destroyed or greatly diminished its agglutinating power. Hence he concluded that the presence of Eberth's bacilli in the circulating blood may account for the absence of a serum reaction therein in certain cases.

PART III.

CHAPTER XI.

SERUM DIAGNOSIS IN OTHER DISEASES.

GLANDERS.

IN December, 1896 (*Journal of Comparative Pathology*, p. 322), McFadyean first observed that the serum of horses affected with glanders would agglutinate glanders bacilli markedly within an hour at a dilution of 1:20. The serum of normal horses had no effect on the bacilli. McFadyean believed, however, that since the mallein test was more useful owing to its earlier appearance, the serum test would be of value chiefly in recognizing the disease post mortem.

Fullerton (*Lancet*, May 1st, 1897) took up the problem of *human* glanders. In a stable boy affected with glanders he found a very prompt reaction at 1:10, while at 1:20 the clumping appeared within ten minutes. By the macroscopic method the test was positive in six hours. The blood of several healthy persons was without any action on the bacilli. But typhoid serum (four cases) did clump the glanders bacilli and diphtheria antitoxin did the same. Moreover, there was found to be some agglutination of the Eberth bacillus by serum from the glanders patient. Fullerton leaves unanswered the question whether by quantitative tests a greater susceptibility of the glanders bacilli to their own serum than to typhoid serum can be demonstrated.

Bourge and Méry (*La Semaine médicale*, 1898, p. 61) experimented on guinea-pigs inoculated with glanders and got an obvious reaction with their serum after nine days (dilution 1:10, after three hours in the thermostat). The clumps were very small, consisting of only three to eight bacilli in each and leaving many organisms freely movable outside. Yet they seemed characteristic—that is, very different from anything that could be obtained with normal serum.

Two horses affected with acute glanders reacted at 1:1,000 and 1:2,000 respectively. Another horse with chronic glanders and a fourth in an acute exacerbation of a chronic affection showed a potency of 1:1,000 in their sera. These results were all obtained by the microscopic test at room temperature and within half an hour.

The serum of sound horses has no clumping action on the glanders bacilli. The serum of horses immunized against diphtheria or against tetanus often clumped the glanders bacilli at 1:50 or higher and "un cheval fébricitant" showed reaction at dilutions as high as 1:300. The writers propose 1:500 as the proper standard test for glanders in horses. Reactions obtained with less diluted serum they consider unreliable. The value of the serum test seems to be chiefly in the most acute cases where the mallein test is usually absent or masked. In chronic cases the serum test is to be considered only as a means of confirming the mallein test.

MALTA FEVER.

The application of the serum diagnosis to Malta fever is one of the most valuable results of Widal's discovery, for on the one hand it bids fair to be the mainstay of our diagnosis between Malta fever and typhoid—diseases wellnigh undistinguishable in many cases hitherto—while on the other hand the identification of Bruce's *micrococcus melitensis* as the organism of Malta fever is greatly advanced. Bruce's description of the organism in 1893 (*Annales de l'Institut Pasteur,* April) had not won him general recognition for the coccus.

Wright's investigations in the application of the Widal reaction to the diagnosis of Malta fever were published in the *British Medical Journal* for January 16th, 1897, and in the *Lancet* (Vol. I., No. 10, 1897).

He used the sedimentation tubes invented by him (see page 25) and obtained the results shown in the table on page 119.

The first eight of these cases had been already diagnosed as Malta fever previous to the serum test. Cases 9, 10, and 11 had been considered typhoid and the others malaria, but the blood of Cases 9, 10, and 11 had no agglutinating effect on the typhoid bacillus and no malarial organisms were demonstrated in Cases 12, 13, and 14.

By the use of this method, Wright has been able to suggest that in all probability the disease "Malta fever" is not confined to the Mediterranean, but exists also in India. At Sabathu the disease appears to be prevalent, also at Hong-Kong. Wright has obtained the reaction in the blood of soldiers who have served only in India.

No.	Period since Infection.	1:10	1:25	1:50	1:100	1:200	1:300	1:1000	Remarks.
1	5 months	+	+	+	0	0	0	0	Higher dilutions not tested.
2	6 "	+	+	+	..	..	..	..	
3	6 "	+	+	+	+	+			
4	8 "	+	+	+	+	+			
5	20 "	+	+	+	+	+			
6	21 "	+	+	+	+	+	+		
7	36 "	+	+	+	0	0			
8	Now sick	+	+	+	+	+	+	+	
9	9 months	+	+	+	+	+			
10	7 "	+	+	+	+	+			No higher dilutions tried.
11	7 "	+	..	..	..	..	..	..	
12	11 "	+	+	+	+	+			
13	6 "	+	+						
14	Now sick	+	+	+	+	+	+	+	

Kretz (*Lancet*, January 22d, 1898) reports a single case of Malta fever in which the serum reacted at 1:1,000 with Bruce's micrococcus melitensis. The serum of this patient had no agglutinating effect on the typhoid bacillus nor on various other organisms experimented with, and the sera of other infections showed no action of the culture of Bruce's coccus used in the tests.

Aldridge (*Lancet*, May 21st, 1898) tested the blood in thirty cases, all of which reacted almost instantaneously with a dilution of 1:10 and within half an hour at 1:100. The reaction first appeared on the fifth day. Four cases of typhoid showed no reaction with the micrococcus melitensis, but promptly agglutinated Eberth's bacilli.

YELLOW FEVER.

P. E. Archinard (*Philadelphia Medical Journal*, Vol. I., No. 6, 1898) states, as the result of his researches as to action of the dried blood of fifty cases of yellow fever on the bacillus icteroides, that from seventy-five to eighty per cent of all cases of yellow fever react positively at some time; thirty of his specimens reacted in a dilution of 1:40 with the bacillus icteroides and not with Eberth's bacillus, even at dilutions as low as 1:5. The remaining twenty specimens reacted but feebly with the bacillus icteroides (1:5) and had an equal effect on the typhoid bacillus.

Pothier tested one hundred and fifty-four cases (*Journal of the American Medical Association*, April 16th, 1898) of yellow fever and did not obtain a single positive reaction on typhoid bacilli. He used Johnston's modification of Widal's method—*i.e.*, one part of dissolved dried blood to ten parts of an attenuated pure culture of typhoid bacilli with a time limit of fifteen minutes. The only positive reactions were in cases which turned out to be typhoid. The blood of twelve cases was tested on the bacillus icteroides. Eight of these showed slight and very slow clumping, usually without loss of motility.

This is very much what Sanarelli himself found (*Annales de l'Institut Pasteur*, October 27th, 1897). According to him the serum of living patients is very slow in its agglutinating action on the bacillus icteroides. After death he found the serum much more potent, but very variable. The serum of dogs immunized against the bacillus icteroides clumps it with lightning speed. Antidiphtheritic serum also agglutinates Sanarelli's bacillus rapidly; antityphoid serum is slower in its action, and "anti-coli" serum has no action.

Otto Lerch (*Journal of the American Medical Association*, February 26th, 1898) publishes what is, I think, the most striking case of agglutination of Sanarelli bacillus by human serum. A nurse in a New Orleans yellow-fever hospital was attacked by the disease in a very typical form. On the second day of his attack the test with Sanarelli bacilli was complete within a few minutes at 1:10, and within twenty minutes at 1:40. The culture was one obtained from Sanarelli himself. In this case the test was of decided diagnostic value, appearing so early as it did in the course of the disease.

Cholera.

The cholera vibrio, being a highly motile organism and one easily cultivated, lends itself readily to the uses of serum diagnosis. In September, 1896, a few months after Widal's application of the phenomenon of agglutination to the diagnosis of typhoid, Achard and Bensaude[1] did the same for cholera. From Cairo and Alexandria they obtained the serum of fourteen cholera patients. Their results are shown in the following table:

Case No.	Age.	Sex.	Day of Disease.	Type of Disease.	Outcome of Case.	Agglutination.
1	30	Male	1st	Severe	Death on 2d day	Present.
2	10	Female	1st	Severe	Death on 3d day	Present.
3	32	Male	2d	Severe	Death on 3d day	Present.
4	12	Male	2d	Severe	Death on 3d day	Present.
5	20	Female	2d	Severe	Unknown	Present.
6	28	Female	3d	Mild	Recovery	Present and very prompt.
7	18	Female	3d	Severe	Death on same day	Absent.
8	30	Female	3d	Severe	Death on 4th day	Absent.
9	35	Male	8d	Mild	Still alive on 5th day	Absent.
10	14	Male	3d	Mild	Still alive on 5th day	Absent.
11	20	Female	4th	Severe	Death on 5th day	Present and very prompt.
12	45	Male	4th	Severe	Death on 5th day	Present and very prompt.
13	36	Male	6th	Moderate	Still alive on 8th day	Present and very prompt.
14	34	Female	35th	Moderate	Recovery on 9th day	Present and very prompt.

In some of these cases the blood had become putrid before it reached Bensaude's laboratory; but in this as in other diseases decomposition proved to be no bar to the activity of the agglutinating substances (see above, page 58).

In the absence of satisfactory clinical reports of these cases it cannot be said with confidence that the reaction is really present on the *first day* of the disease (see Cases 1 and 2). In cholera as in typhoid it is difficult to fix the first day of the illness and to separate it from the more or less indefinite prodromes. The report "first day" means different things to different observers, and we have no means of knowing how it was reckoned by the Egyptian physicians from whom the blood was obtained.

As will be seen from the table, the reaction was absent in only a single case, one in which death took place on the same day that the specimen of blood was obtained. The occasional failure of the reaction in rapidly fatal cases of typhoid has been already mentioned.

The dilution used in these cases was as a rule 1:10, but most of them reacted promptly at 1:20, and Case 14—in which the blood was obtained in convalescence—showed the reaction in a dilution of 1:120. The time limit was one hour; but most cases reacted within ten minutes.

Achard and Bensaude used as a control the serum of thirty

[1] Achard and Bensaude: Press médicale, September 26th, 1896; Société méd. des Hôpitaux, April 23d, 1897.

other cases, six of them normal and the remainder suffering from a variety of diseases (pleurisy, pneumonia, chronic bronchitis with emphysema, cancer of the liver, cirrhosis of the liver, Bright's disease, meningitis, typhoid, puerperal fever, diabetes, etc.).

Mixed with the culture of cholera vibrios the serum of these cases showed no clumping except in two uræmic cases in which a few small clumps formed. These were much smaller than those formed by the cholera serum and occurred chiefly in dilutions less than 1:10. By using a dilution of 1:15 they were in all cases avoided.

These results are not in accord with those of Pfeiffer and Kolle, who obtained with normal human serum a clumping of the cholera vibrios even when diluted 1:20.

Bensaude points out that if bouillon cultures are used in the test one must take care to select only such as have no pellicle—for a fragment of the pellicle looks almost exactly like the clumps of a typical reaction, and such fragments might easily be present in the portion of the culture used for examination and lead to a false report.

This difficulty can be avoided by using suspensions made by rubbing up a loopful of a sixteen to twenty-four hour gelatin culture with sterile bouillon. [The technique of preparing a fresh emulsion has already been described. The emulsion must always be examined microscopically before any serum has been added, to make sure that the vibrios have been effectually dissociated.]

The slow method may also be used and generally shows considerable precipitation within a couple of hours. Sometimes a precipitate visible to the naked eye forms at room temperature within a few minutes. The quick method is, however, more reliable as well as more convenient.

As in typhoid, either immobilization or clumping may occur without the other. A certain amount of degeneration (Pfeiffer's phenomenon) can sometimes be seen in the clumped vibrios.

The reaction can be produced on dead vibrios[1] as in typhoid, and when so produced has the advantage of not being simulated by bits of pellicle; no pellicle is formed. But the vibrios being deprived of their motility, the reaction is less complete and characteristic. In working with dead vibrios Bensaude used a three-

---

[1] Bordet: Annales de l'Institut Pasteur, April, 1896, p. 208.

per-cent solution of fluorite of sodium, which does not coagulate the albuminoids of the serum.

The blood of cholera patients is sometimes difficult to work with on account of its concentrated condition, which renders it hard to collect in the little tubes used for the purpose. Occasionally a patient has to be bled to get enough blood for the test!

The reaction can be carried out by the dried-blood method as in typhoid, using glass or glazed paper to dry the blood on. Bensaude found that blood dried in this way retained its agglutinating power intact after five months.

These researches of Achard and Bensaude show at once that the serum diagnosis is applicable to cholera, and clinically valuable since it can be more easily and quickly carried out than the bacteriological examination of the stools.

The use of serum is also valuable to distinguish the different closely allied species of vibrio. Ten different races of cholera vibrios, from one to twelve years old and coming from various parts of Europe and Asia, all reacted about alike with a given cholera serum, which, however, was without action on the vibrio Metchnikovi, the vibrio Finkler, and the vibrio Massaouah.

Blackstein's discovery that chrysoidin would also clump cholera vibrios when added to the bouillon culture in the proportion of ten drops of the twenty-five-per-cent solution to 3 c.c. of culture, seemed at first to provide a test for identifying the organism of cholera even more certain and available than the serum test. But Bensaude, who controlled Blackstein's results, found the action of chrysoidin very inconstant and unreliable, by no means fitted to take the place of cholera serum.

Engel (*Centralblatt für Bakteriologie*, January 30th, 1897) has shown that the action of chrysoidin varies according to the source from which it is obtained and agglutinates the vibrio Finkler fully as well as the cholera vibrio.

So far no sufficient trial of the serum diagnosis of cholera has been made at the site of the epidemic, and we do not yet know whether its value is great or small, how early it usually gives data, and whether it will help to clear up the relationship between Asiatic cholera and cholera nostras.

The Bubonic Plague.

The German commission sent to Bombay to investigate the whole subject of the plague first discovered the applicability of the method of serum diagnosis to this disease ("Mittheilungen der deutschen Pestkommission aus Bombay." *Deut. med. Woch.*, 1897, No. 17). The commissioners reported that the serum of convalescents and of animals immunized against the plague had a marked agglutinating action on plague bacilli. As to its practical value they made no report.

Paltauf (*La Semaine médicale*, June 2d, 1897) found the reaction present in the blood of guinea-pigs killed by inoculation of dead plague bacilli, but failed to get it in the blood of an immunized sheep and in the pus of an abscess produced in a horse by injections of plague bacilli. The serum of this horse, however, and of others subjected to intravenous inoculation were found markedly agglutinative.

Wyssokowitch and Zobobati (*Archives Russes de Pathol.*, May 31st, 1897) state that the clumping power first appears in the blood on the fifth, sixth, or seventh day after the onset of symptoms, gradually increases up to the time of the fourth week of the disease, and then diminishes. They found it most marked in the most serious cases. In patients dying of very acute forms of the disease and in the pneumonic forms of infection the reaction did not appear. Post mortem they found it absent.

Zabolotmy, reporting on forty cases (*Deut. med. Woch.*, 1897, p. 392), arrives at practically the same results. He got no reactions within the first week of the disease, positive reactions at $1:10$ in the second week, and at $1:50$ in convalescence (third and fourth weeks). The capsules of the organism are removed or made invisible by the action of agglutinating serum. These researches were carried out in Bombay during the epidemic of 1897. Stricker (*Münch. med. Woch.*, 1898, No. 7) states that in mild or abortive cases the reaction does not occur.

It does not seem, so far, as if the serum test were to be of much practical value in the diagnosis of the plague.

Pneumococcus Infections.

In one of his earliest communications Widal stated that he had attempted to apply the serum diagnosis to cases of pneumonia and had been unsuccessful. This for a time discouraged further investigation along this line, but it was not long before attempts were renewed, for the results of older investigations gave reason for believing that Widal had been unfortunate in his experiments. As far back as 1891 Metchnikoff had noticed (*Annales de l'Institut Pasteur*, 1891, p. 374) that when pneumococci were grown in the serum of rabbits immunized against this coccus, long chains and masses were formed, unlike those seen when the organism was cultivated in normal serum. In 1892 Mosny reported that no turbidity was produced when pneumococci were grown in immune rabbits' serum (*Arch. de Méd. expérimentale*, 1892, p. 228). In the following year Issaeff confirmed these observations (*Annales de l'Institut Pasteur*, 1893, p. 269), and noted that besides producing no turbidity the pneumococci were deposited in masses at the bottom of a tube of immune serum at the end of twenty-four hours.

Washburn (*Journal of Path. and Bact.*, 1895, Vol. III.) observed the same facts and naïvely commented on this " peculiar way of growing" without suspecting that it could have any practical application. He showed that the sediment deposited at the bottom of tubes of immune serum in which pneumococci had been planted was composed of masses of pneumococci which, however, were still capable of reacting well to staining reagents.

Bezançon and Griffon (Soc. de Biol., June 5th and 19th, 1897) were the first to investigate the serum of animals during the period of infection before immunity had been produced. Their experiments were difficult to carry out successfully on account of the great susceptibility of the animals to pneumococcus infection and the rapidly fatal effect of the injections. They found, however, that the blood of mice and rabbits killed by pneumococcus injections possessed agglutinative power over pneumococci, and possessed it to a much greater extent than the blood of vaccinated animals. The precipitate formed after twenty-four hours at the bottom of a tube of serum from an infected animal was so tough and consistent that shaking had no effect on it and it was hard to pull apart. "All the pneumococci were precipitated and plastered together into a wad. Under the microscope the organisms appeared in large heaps which are quite different from those seen in the precipitate caused by the serum of immunized animals (Fig. 5), and entirely without capsules." When grown in ordinary serum the pneumococci retain their capsules and are entirely separate one from another (Fig. 6). This agglutinative power in the serum of infected rabbits and mice was strong enough to act even when diluted 1:50, and is to be seen macroscopically after twenty-four hours at 37° C.

In man similar experiments have proved successful in the hands of Bezançon and Griffon—though the results were far less striking. Pneumococci grown in serum taken from a vein at the

 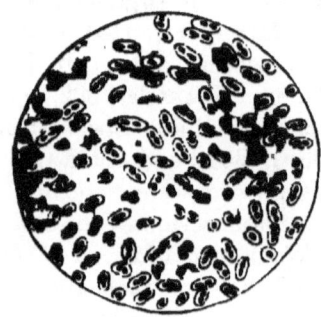

FIG. 5.—Pneumococci Cultivated in Serum of Immunized Rabbit.

FIG. 6.—Pneumococcus Cultivated in Normal Human Serum.

bend of the elbow in seven cases of pneumococcus infection were clumped in every case. Five of these were cases of pneumonia, one was a case of meningitis, and one a case of empyema. The tubes of serum taken from these patients and inoculated with pneumococci remained entirely free from turbidity after twenty-four hours in the thermostat, and at the bottom were found very small but numerous clumps (see Fig. 7) and chains of various lengths, often twisted into shapes resembling the letters O, S, or U. No capsules were distinguishable. There were no isolated organisms.

In healthy serum or serum taken from cases of other diseases (ten individuals in all were tested) no such groups or chains were found. The growth was very slight both in the healthy and in the infected serum and soon died out.

In four of the seven cases above described the reaction was

found on the sixth or seventh day, in one as early as the second day. In one case tested a second time on the seventh day after defervescence the serum had partly lost its clumping power. Between the clumps there were some isolated organisms.

In two other cases examined by Bezançon and Griffon the serum agglutinated only the organism which caused the special case under examination and had no action on other races of pneumococci. Apparently then the serum reaction can be used not only for diagnosis of pneumococcus infections but as a means of differentiating closely allied species of diplococci.

In a later communication (*La Semaine médicale*, 1898, p. 178) Bezançon reports twelve more cases, all successful. As before he found *either* clumps or long chains.[1] Sometimes the serum reacted only with the organism isolated from that particular case, as in the group of cases previously reported.

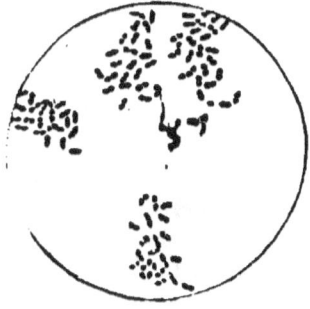

FIG. 7.—Pneumococci Cultivated in Serum of a Case of Pneumococcus Infection in Man.

When the organisms get into the general circulation the serum reaction becomes much less marked.

I know of no other investigations on this subject except those of Block (personal communication), who was unable to get any satisfactory results.

COLON-BACILLUS INFECTIONS.

In view of the frequent association of the colon-bacillus with various diseases and of our ignorance as regards its pathogenic properties, it seems of great importance that further evidence should be obtained if possible through serum diagnosis, but as yet investigations in this direction have not been fruitful. Grüber and Durham showed in 1896 that the serum of animals immunized against the bacillus coli communis would clump the race of bacilli by which the immunity was secured and would not clump typhoid bacilli unless it was unusually potent. Similarly they found no action of typhoid serum on colon bacilli

[1] Compare McWeeney's observations on typhoid, p. 20.

except when the typhoid serum was very concentrated. These observations have been contradicted by others, though not always intelligently.

Many races of colon bacilli clump spontaneously under certain conditions, or, if not spontaneously, it may need only the addition of any serum, normal or diseased, to produce clumping. Thus Widal (*La Semaine médicale*, 1896, p. 312) found on testing twenty cases of supposed colon-bacillus infection (cystitis of young girls, appendicitis, etc.) that their serum had no more effect on colon bacilli than the serum of other diseases. In some cases of typhoid in which a secondary colon-bacillus infection was suspected, he thought that the serum was perhaps more agglutinative than other sera. Chantemesse (*La Semaine médicale*, 1896, p. 301) noticed no especial effect of typhoid serum on colon bacilli. Courmont (Soc. de Biol., July 28th, 1896) states that typhoid serum does clump bacillus coli, and various other organisms, though to a much less extent than typhoid bacilli. He does not state whether the action of typhoid serum on colon bacilli is any greater than that of other sera.

Vedel (*La Semaine médicale*, 1896, p. 312) also reports some action of typhoid serum on young cultures of colon bacilli and describes one case, clinically a very mild typhoid, in which there was no reaction with typhoid bacilli but a marked one with colon bacilli.

Rodet (Soc. de Biol., July 31st, 1896) immunized two sheep against typhoid and against colon infection respectively and found that while each bacillus was strongly agglutinated by its homologous serum, the effect of the typhoid serum on the colon race was not noticeable until a dilution of 1:2 was reached. Fresh sheep serum has a slight effect on colon bacilli even at 1:40.

Achard got negative results in three cases of colon-bacillus infection (two of them pyelonephritis and one pleurisy), although in each case the organism isolated from the patient was used to test his serum.

Bensaude experimented with eight guinea-pigs, and found that it was difficult to produce any agglutinative power in their serum by inoculating them with colon bacilli. Five of them developed absolutely no clumping power even with the same race of bacilli used to inoculate them; after eight inoculations the sera of three of the pigs clumped certain cultures of colon bacilli, but had no effect on the others. Yet the doses inoculated were large enough to cause death in some of the animals.

Rodet (*La Semaine médicale*, 1897, p. 361) found that the serum of animals immunized against bacillus coli seemed to clump typhoid bacilli as strongly as it did bacillus coli and as strongly as the serum of animals immunized against typhoid. He concludes that we had better try bacillus coli serum to cure typhoid.

Van de Velde (*Bull. de l'Acad. de Belgique*, March 27th, 1897) immunized a horse by successive inoculations for over a year and then tested his serum on twenty-five different races of colon bacilli. Only four of them showed any reaction when diluted 1:10.

Against these discouraging results we have to balance off the following:

Johnston (*Montreal Medical Journal*, March, 1897) states that he has found "reactions with the colon bacillus to be rare with typhoid blood or serum (even in cases when perforative peritonitis had occurred), *provided* the typhoid reaction was well marked. On the other hand, he has been struck by the large proportion of positive colon reactions in cases having stepladder temperature and other symptoms strongly resembling typhoid but without the typhoid reactions.

"We think that under these circumstances the colon reaction may have a real diagnostic importance, and indicates that the colon infection, whether occurring alone or as a secondary complication of typhoid, may be playing an important part in the production of the patient's condition. The whole question of associated colon infection deserves further study.

"The reaction can be tested with ease by placing a duplicate drop of blood solution or serum on the cover slip with the drop to be tested by typhoid culture and mixing it with a drop of colon-bacillus culture. Pseudo-reactions can be avoided by using stock cultures kept at room temperature, and transplanted infrequently. Test cultures grown in bouillon from the stock at room temperature for twenty-four hours are free from scum or sediment, and give reliable results. The conflicting results just mentioned may have been due to pseudo-reactions having been taken seriously.

"In our case of apparently genuine typhoid without serum reaction (in which, by the way, the test was first applied during the third week) the blood reacted very decidedly to B. coli, producing typical clumping. The same held good of four other blood samples referred to us for examination as having a clinical course like typhoid, but with negative serum reaction. A complete colon reaction we have found to be exceptional in ordinary typhoid, and its presence would indicate a condition of coli intoxication sufficient to explain the existence of many symptoms giving to typhoid its ordinary clinical features. Whether this excludes typhoid is another question. W. H. Park has observed a case of fever with no typhoid serum reaction, where he was able to cultivate the typhoid bacillus by spleen puncture. Later on in the case, however, a relapse occurred and the reaction appeared. The possibility of a latent typhoid infection overshadowed by toxic phenomena, due to concurrent action of the colon bacillus, is quite consistent with the generally accepted opinion that many of the symptoms in typhoid, and especially the intestinal ones, are due to secondary infection by B. coli. It follows that in severe cases of typhoid type, with no typhoid reaction, the blood should be tested with a culture of B. coli and a bacteriological study made by examination of the stools or by spleen puncture."

Stern states clearly (*Centralb. für Bakteriol.*, April 30th, 1898) that many races of colon bacilli have a tendency to spontaneous agglutination. He also experimented with twenty-five sera taken from healthy persons or from patients with various diseases, and not infrequently got clumping within two hours in dilutions of 1:25 or even 1:60. Among normal sera he found some that clump colon bacilli better than typhoid bacilli, and others which had greater power over the latter than over the former. In typhoid fever he believes, as Johnston does (*vide supra*), that there is sometimes a secondary infection with colon bacilli. As a rule he finds that typhoid sera clump typhoid bacilli more strongly than they do colon bacilli. For example, one serum which agglutinated typhoid bacilli at a dilution of 1:14,000 had no effect on colon bacilli when the dilution was greater than 1:100. Yet in his experience he had met with five cases of typhoid fever in which the serum was more potent with colon bacilli than with typhoid bacilli. For example, in one case the serum would not react above 1:140 with typhoid bacilli, while with colon bacilli it produced a positive reaction at 1:250. [With this case we may compare that reported by Widal and Noblecourt (*La Semaine médicale*, August 4th, 1897):

a typhoid patient whose serum clumped typhoid bacilli only 1:20, but was potent with a race of "paracolon bacilli" at 1:12,000. Compare also Durham's conclusions on page 111.]

Pfaundler (*Centralb. für Bakt.*, Vol. 115, No. 4) examined the blood in eight cases of what he deemed colon-bacillus infection. He got no reactions except when he used the particular race of bacilli which had already been isolated from the case under investigation, and then only while fever was present. He used the macroscopic method in all cases and observed his reactions chiefly after twenty-four hours and with "nascent bacilli."

Widal returned to the study of colon-bacillus reactions in 1897 and with better success. In *La Semaine médicale* (page 285) he reports a case of phthisis in which an abscess developed near the œsophagus. From this abscess was isolated a bacillus resembling the colon group, except that it produced no indol and did not ferment lactose. On the fourth day after operation this organism (which Widal calls a "paracolon bacillus") was agglutinated by the patient's serum even after dilution at 1:1,000. With ordinary races of colon bacilli the patient's blood showed no reaction, and typhoid serum showed no clumping with the "paracolon bacillus."

Soon after this Lesage reported in the same journal (page 383, 1897) an epidemic of enteritis in infants in forty out of fifty of whom the organism isolated from the stools was very strongly clumped by the sera of the infant from whom the organism was obtained as well as by the serum of each of the others. Various races of colon bacilli isolated from the intestines of normal infants of the same age and from adults, were not at all affected by the serum of any of these infants, nor did the serum of healthy infants show any effect upon the organism of this epidemic. Typhoid serum had no effect on this organism, nor were typhoid bacilli clumped by the serum of the sick babies.

The duration of the reaction was short, and in cases which became chronic it soon ceased. In cases which were fatal after the reaction had ceased to be demonstrable it could be found post mortem in the blood from the liver, but not elsewhere.

The reports of Lesage, Widal, Stern, and Johnston certainly show that the subject of colon-bacillus infections is worth further study, but the evidence is so contradictory that I do not feel able to draw from it any definite conclusions. I have no personal experience in the matter.

TUBERCULOSIS.

Arloing (*La Semaine médicale*, 1898, No. 20) has succeeded in getting tubercle bacilli to grow in liquid media so as to produce a diffuse turbidity due to the isolated organisms. He achieved this by growing the bacilli on cooked potato constantly saturated with glycerin and water. Under these conditions there forms a growth which can be dissociated in glycerin bouillon and will grow there for some time.

With a culture so obtained Arloing tried first the blood of animals which had been inoculated either with tuberculin or with live bacilli, and got clumping in each case. One hundred cases of pulmonary tuberculosis were then tried and 94 were found to react positively. The agglutination was complete in 57 and incomplete in 37.

Out of 100 cases of surgical tuberculosis 91 reacted more or less—35 of them completely and 56 incompletely.

In 100 cases of other diseases Arloing got 32 positive reactions, 11 of which were complete and 21 incomplete. Some of these may have had tuberculosis, he thinks.

Finally 100 healthy persons were tested, 22 of whom showed positive reactions, 11 complete and 11 incomplete.

Courmont tested the clumping power of peritoneal, synovial, and pleural effusions in tuberculous and non-tuberculous cases. In 10 out of 11 cases of tuberculous pleurisy the effusion clumped tubercle bacilli in dilutions varying from 1:10 up to 1:20. Normal or non-tuberculous pleural fluid never agglutinated the bacilli in dilutions higher than 1:5. Out of 9 cases of pleuritic effusion *a frigore* he found clumping in 4. Five cases of tuberculous peritonitis and 8 cases of tuberculous synovitis showed positive results. Eight cases of non-tuberculous ascites were negative.

STREPTOCOCCUS INFECTIONS.

Bensaude, working under Widal's direction, tested ten patients suffering from various types of streptococcus infection (erysipelas, puerperal sepsis, general streptococcus sepsis with metastases, streptococcus abscess of the leg, etc.). The blood of two of these patients showed absolutely typical clumping with bouillon culture of streptococci, the tube in which they were mixed becoming entirely clear and depositing masses of cocci at the bottom after some hours in the thermostat. The other eight cases and twelve normal ones used as controls showed no clumping.

Microscopic examination of the precipitate in the positive cases showed it to be composed of masses not at all like those formed in ordinary streptococcus cultures. They were made up of "a great number of highly refractile, isolated granules looking like staphylococci or like cholera vibrios in Pfeiffer's phenomenon." No chains were visible, but in the centre of the heaps there were granules larger and more refractile than any of the original cocci.

Cultures grown at laboratory temperature were found better suited to the test than those grown in the thermostat.

A second set of experiments, made this time with "nascent bacilli" (see above, page 21), were no more successful than the first.

Marmorek's antistreptococcus serum and various other protective sera were also tried by Bensaude on six cultures of streptococci. His results are shown in the following table:

Culture No.	Antistreptococcus Serum.		Antidiphtheritic Serum.		Antitetanus Serum.		Antityphoid Serum.		Normal Serum.	
	After Half Hour.	After 18 Hours.	After Half Hour.	After 18 Hours.	After Half Hour.	After 18 Hours.	After Half Hour.	After 18 Hours.	After Half Hour.	After 18 Hours.
I.	+	+	+	+	+	+	?	+	+	+
II.	0	0	0	0	0	0	0	0	0	0
III.	0	+	0	0	0	0	0	0	0	+
IV.	0	0	0	+	0	+	0	0	0	0
V.	0	0	0	0	0	0	0	0	0	0
VI.	0	0	0	+	0	0	0	0	0	0

From this table it would appear that it depends rather on the particular culture used than on the kind of serum whether or not agglutination shall occur.

These results have been confirmed by v. de Velde, who studied the effects of four sera from horses immunized against the streptococcus on fifteen different races of streptococci and found no constant agglutinative action.

On the other hand, Durham (*Lancet*, December 19th, 1896) speaks of Marmorek's serum as having a marked clumping action on streptococci, and Masius (*La Semaine médicale*, 1897, p. 114) found that this serum gave positive reactions with four races of this organism.

LEPROSY.

Spronck,[1] of Utrecht, has recently succeeded in cultivating from the tissues of leprous patients a bacillus apparently identical with Hansen's leprosy bacillus. After cultivation on neutral glycerinated potato for ten days at 38° C. it can be transferred to fish bouillon and used in this medium for serum-diagnosis.

The blood of 68 non-leprous cases never produced agglutination of this bacillus in dilutions of 1:30 or higher, and in only 2 of the 68 cases could clumping be produced at 1:20.

The blood of 27 non-leprous cadavers showed a slightly greater clumping power, being active at 1:40 in one case, but never over 1:50.

Twelve cases of leprosy were tested, with the results shown in the following table:

Case.	Form.	Duration.	Agglutination.		Remarks.
1	Mixed	5 years.	1:1,000 and	1:1,100	Bacilli of Hansen found in tissues.
2	"	10 "	1:200 "	1:300	Advanced case with great mutilation.
3	"	7 "	1:260 "	1:300	Bacilli in tissues.
4	"	7 "	1:70 "	1:80	" " "
5	"	10 "	1:300 "	1:400	
6	"	15 "	1:70 "	1:80	
7	"	11 "	1:60 "	1:70	" " "
8	Anæsthetic	15 "	1:30 "	1:40	" " "
9	"	15 "	1:20 "	1:30	
10	"	32 "	1:20 "	1:30	" " "
11	Anæsthetic with leprous exanthem resembling eczema marginatum.	10 "	1:200 "	1:300	
12	Tuberculous.	2 "	1:60 "	1:70	" " "

From these cases it appears that the quiescent, anæsthetic forms without considerable mutilation possess far less agglutinative power than the mixed and active forms, but that even the anæsthetic cases possess a greater agglutinative power than any of the non-leprous cases so far experimented with.

In case 11 of this series Spronck found the serum test of great diagnostic value, since the lesions were very atypical and

[1] Spronck : La Semaine médicale, September 28th, 1898.

the diagnosis was uncertain. Hansen's bacilli were later found in the tissues, confirming the results of the serum test.

The leprous serum retains its clumping power six weeks or more. A one-per-cent phenicated bouillon is added and the light excluded. Dried serum retains this clumping power at least a month.

DIPHTHERIA.

Delépine (*Medical Chronicle*, October, 1896) reported that the Klebs-Loeffler bacillus was distinctly agglutinated by antidiphtheritic serum. He does not mention the action of other sera on it.

Nicolas and Charrin (*La Semaine médicale*, December 9th, 1896) confirmed these results, and noticed also that the virulence of the agglutinated bacilli was diminished by the process.

Later (*La Semaine médicale*, 1897, p. 37) Nicolas tested post mortem the serum of a patient who had just died of diphtheria and who had received no antitoxin. This serum had no effect either on fully developed or on nascent Klebs-Loeffler bacilli. A second specimen of serum taken during life on the fourth day of the disease from a patient who had never had antitoxin was likewise without any action on diphtheria bacilli. On the other hand, he found that on the day following the injection of antitoxin (20 c.c.) the serum distinctly agglutinated Klebs-Loeffler bacilli, especially the "nascent bacilli" (see above, page 21). The power persists about two weeks, gradually diminishing after the first day following the injection. Nicolas finds that cultures work decidedly better than emulsions.

Experimenting with antidiphtheritic serum he observed that three drops of a serum having an immunizing power above 1 : 50,000 added to 1 c.c. of bouillon culture of Klebs-Loeffler bacilli soon produce a fine flocculent precipitate and leave the upper portion of the bouillon clear. This phenomenon begins to show itself in from twenty minutes to two hours, and is completed in from eighteen to twenty-four hours. If ten drops of antitoxin are added to 1 c.c. of culture, the precipitate falls within a few minutes. The microscope always shows this precipitate to be made up of deformed and agglomerated bacilli.

Tetanus.

The tetanus bacillus, as was first shown by Bordet (*Annales de l'Institut Pasteur*, 1896, p. 211), is strongly clumped by the serum of normal horses. With antitetanus horse serum agglutination is instantaneous with a dilution of 1:20.

Bensaude examined four cases of tetanus in man with negative results. The serum was taken on the eighth, tenth, twentieth, and twenty-first days respectively. Three of these had received antitetanus serum. Post-mortem investigation of three cases showed no clumping power.

Weinberg (reference in Bensaude, p. 217) tried the serum of three cases at autopsy, also with negative results.

Sabrazès and Rivière (*Soc. de Biol.*, June 26th, 1897) used cultures in a vacuum and obtained clumping with the serum of a patient on the eighth day of an attack of tetanus. The serum of a tetanized dog likewise reacted, and very rapidly when diluted 1:10 and 1:20.

Normal dog serum and human serum gave no reaction under similar conditions.

Peripneumonia of Cattle.

Arloing noticed as long ago as July 5th, 1896 (*Lyon Médical*) that the serum of animals rendered immune to this disease would clump the *pneumobacillus bovis* in bouillon suspensions. Within fifteen minutes the opacity began to clear, and in less than half an hour the clumps began to pile up at the bottom of the tube. At the end of an hour and a half the upper two-thirds of the bouillon had become entirely clear. In twenty-four hours many of the bacilli had undergone granular degeneration and become difficult to stain.

In the blood of a heifer which had been given 800 c.c. of pleuritic effusion from a case of peripneumonia in the course of five months, the clumping power was found to be 1:200. A normal heifer had little or no clumping power in its blood. Two other inoculated cows showed likewise positive reactions. On other micro-organisms their serum had no effect.

Later (*La Semaine médicale*, 1897, p. 38) Arloing published the results of tests with the body fluids of an animal which had

died of the disease. He found the order of potency to be as follows:
1. Serum.
2. Juice of lymph glands.
3. Bile and liver juice.
4. The splenic juice had no clumping power.

There appeared to Arloing to be less clumping power near the lesions than at a distance from them.

Hog Cholera.

Dawson experimented with the blood of a rabbit which had been inoculated with the bacilli of hog cholera (*New York Medical Journal*, February 20th, 1897) and got prompt clumping of the bacilli with it. A control rabbit's blood was negative.

Later Cashin (*Journal of the American Medical Association*, 1897, p. 784) examined seven hogs suspected of having the disease. In four cases he got a prompt reaction, and the autopsy confirmed the result of the test in all of them. Two of the others which had shown no reaction were proved later not to have hog cholera. In the seventh case the reaction was positive, though the hog had passed through and recovered from his attack four months before. In experimental infections it was shown that the agglutinating power is rapidly acquired.

The diagnosis of hog cholera from the bubonic plague and from anthrax bids fair to be much simpler in the future owing to this test, and as Bensaude suggests it would be a quick way of testing a bit of meat suspected of coming from a cholera-infected hog.

Pictou Cattle Disease.

Johnston and Hammond (*British Medical Journal*, February 5th, 1898) state that the specific organism of the Pictou cattle disease shows a positive agglutinative reaction with the serum of animals suffering from that disease. The blood of healthy cattle had no action on the micro-organism. The dilution used was 1:25.

Proteus Infections.

Achard and Lannelongue (*La Semaine médicale*, 1896, p. 400) reported experiments on animals with the proteus vugaris, proteus mirabilis, and other species. As is well known, proteus infections in the human being are not of very rare occurrence, though they are more frequently found as a terminal infection or post-mortem parasite than as a primary cause of disease. Animals immunized against a certain variety of proteus often

Fig. 8.—Oidium Cultivated in Serum of Subject Immunized Against this Organism.

show in their blood after three or four days a considerable agglutinating power, especially upon the particular race of organisms used in the immunizing process. If the degree of immunity is relatively high, the clumping power ceases to be confined in its effects to the particular race used for inoculation and extends to any culture of the same species. Differences between different species may, so the writers think, be brought out through their reaction or failure to react with a given serum, as has been done with the different allied species of vibrios.

They found, moreover, that it was perfectly possible to immunize an animal against two or more species, thus giving its serum a power to clump any of these species. The agglutinating power is acquired in about three or four days after inoculation, and exists not only in the blood but to a less extent in the urine and in the aqueous humor. In the bile and in the semen it was found absent. It persists after death and even in putrefaction, but never occurs in the fluids of bodies infected after death. This fact is sometimes of importance. If at an autopsy one finds the proteus

Fig. 9.—Oidium Cultivated in Normal Serum.

in the tissue and desires to know whether the individual had been infected with the organism during life, or whether it was a post-mortem invasion, one has only to test the serum on the organism. If it is clumped thereby, the infection must have taken place more than three days before death.

Whether the serum reaction with organisms of the proteus group is destined to be of any practical application for diagnostic purposes, the writers do not feel sure. It does not occur with healthy serum. It is occasionally found in affections involving intestinal ulceration (typhoid, cancer of the gut), but is not constant in such conditions.

Achard (*La Semaine médicale*, 1897, p. 85) succeeded in immunizing animals so powerfully against the proteus that the agglutinating power passed through the placenta to the animal's embryo, and was found present though faint in the new-born young.

Roger, experimenting with the oidium albicans (Soc. de Biol., July 4th, 1896), found that it grew but feebly in the serum of animals immunized against it, but that such organisms as did grow were formed in masses, and fell to the bottom (see Figs. 8 and 9) and ceased to multiply. Charrin and Ostrowsky verified these observations.

No experiments with human thrush have so far been reported.

Staphylococcus Infection.

Achard examined the sera of four patients with staphylococcus osteomyelitis without finding any agglutinating power in them. It is very difficult to obtain growths of staphylococci free from spontaneous clumping.

Anthrax.

The same difficulty is present in the case of anthrax growths. It is almost impossible to get cultures or suspensions in which the bacilli are thoroughly dissociated. Achard tried the serum of two cases of anthrax on cultures of as diffuse a growth as he could obtain, but without any satisfactory results.

Psittacosis.

Psittacosis is an infectious disease not uncommon and very fatal among parrots, and communicable to other animals and to man. Since 1892 it has been supposed to be due to a bacillus very closely akin to that of typhoid fever. This bacillus, first described by Nocard[1] and later studied by Gilbert and Four-

[1] Nocard: "Conseil d'Hygiène publique et de Salubrité," March 24th, 1893.

nier,[1] grows on potato like the colon bacillus and ferments media containing glycose or maltose. It is more virulent than Eberth's bacillus and readily infects animals when given with their food. On the other hand, it differs from the ordinary colon bacillus in that it will not produce indol nor coagulate milk nor ferment lactose. It is also more motile than most races of colon bacilli. It will grow on the denuded surface of media from which cultures of colon bacilli or of Eberth's bacilli have been scraped off, whereas typhoid bacilli will not grow on the surface of a typhoid culture so denuded, nor colon bacilli on a scraped colon-culture tube.

Cases of psittacosis in man are rare, and no very definite symptomatology has yet been made out, but the clinical course seems to resemble that of typhoid very considerably.

The facts as to the possibility of a serum diagnosis of the disease are somewhat contradictory and inconclusive.

One very interesting case is related by Bensaude. I quote from his own account: "A woman of twenty-four came under observation while suffering from an acute febrile affection somewhat resembling typhoid fever but without any mental dulness or rose spots. A pyelonephritis was also present, from the pus of which there was isolated an organism having the characteristics of the psittacosis bacillus."

The serum of this patient clumped cultures of the psittacosis bacillus very strongly (1:40) and had only a very slight effect on four out of fourteen races of typhoid bacilli on which it was tried (1:10). The serum of animals inoculated with the psittacosis bacillus behaves in just this way—that is, it clumps the psittacosis bacillus very strongly, and has a feeble effect on *some* typhoid cultures.

Gilbert and Fournier mention another successful case (Soc. de Biol., December 25th, 1896). In five other cases of supposed psittacosis no characteristic reaction was obtained. In all of these cases the diagnosis was based upon the close association with parrots which had lately sickened and died.

The number of cases reported so far is too small to enable us to judge as to its clinical value.

[1] Gilbert et Fournier: Bull. de l'Académie de Méd., October 20th, 1896; Soc. de Biologie, December 12th, 1896; Presse médicale, January 16th, 1897.

Relapsing Fever.

The serum diagnosis of relapsing fever so ingeniously worked out by Löwenthal (*Deut. med. Woch.*, 1897, No. 35, and Moscow Congress, August 23d, 1897) does not come under the head of an agglutination reaction, but is more akin to Pfeiffer's phenomenon. It is due rather to an antibacterial than to a clumping power on the serum, but its practical value for diagnosis is apparently very considerable in the countries where epidemics of relapsing fever are common.

As is well known, the diagnosis of the disease during the febrile paroxysms is exceedingly easy, since the specific spirochæte which causes the symptoms is then always present in the peripheral circulation and readily seen on microscopical examination of a fresh slide-and-cover-glass specimen. But, on the other hand, the diagnosis during the afebrile interval has hitherto been a matter of the greatest difficulty, since the organisms then disappear entirely from the peripheral circulation and leave behind no trace of their presence that has hitherto been recognizable. Patients first seen at the end of a febrile attack which may have been one of relapsing fever are very anxious to know whether they are to expect another attack.

The method practised by Löwenthal can be carried out only in case we can secure from another patient a drop of fresh blood containing live spirochætes, for we know of no way of cultivating the organism outside the body. Since, however, the disease occurs almost invariably in epidemics, it is nearly always possible to get blood containing the organisms.

The method of testing cases between attacks is as follows: A drop of the suspected blood is mixed with one known to contain living spirochætes, and the mixture is sealed up with wax between slide and cover-glass and left for half an hour in the thermostat at 37°. If the suspected blood be really from a patient who has recently passed through a paroxysm of the disease, its bactericidal power is exerted upon the spirochætes so that they lose their motility and their characteristic spiral curl, and accumulate in bunches. The blood of other diseases has no such effect on the organisms. Löwenthal tried blood from pneumonia, influenza, typhoid, and rheumatism (three cases each) and from malaria, all with negative results.

Thirty-nine cases suspected of being relapsing fever in the apyretic interval were tested by Löwenthal. Thirty showed positive reactions, and all turned out to have relapsing fever, as was proved by the occurrence of a second paroxysm with spirochætes in the blood. The other nine cases showed no effect on the spirochætes and turned out to have other diseases.

Just before the next paroxysm is to occur the bactericidal power of the blood dies out.

The reaction was found to occur even in the abortive and very mild cases with very few spirochætes in the blood.

Serum Prognosis of Relapsing Fever.

As this reaction expresses a direct bactericidal power in the blood, it can be used for prognosis as well as for diagnosis in a way that is impossible in the case of the more subtle processes involved in the agglutination test. Löwenthal at once availed himself of this fact (*Deut. med. Woch.*, September 16th, 1897). He found that by measuring the amount of specific bactericidal power present in a given case during the apyretic period he could predict whether a second attack was to follow or whether the force of the disease had spent itself in the first conflict with the body. By experiments in over one hundred cases he found that if the power to immobilize the live spirochæte of another case *within one hour* persisted later than the seventh day after the last paroxysm, no second attack followed. If it needed more than an hour to accomplish the result, the patient was doomed to a relapse unless this could be prevented by the use of antispirilla serum.

Summing up the practical uses of Löwenthal's discovery, it appears that it enables us:

1. To diagnose the disease between attacks, and so very possibly prevent a second attack by serum therapy.
2. To recognize abortive cases in which the organisms are very scanty in the peripheral blood.
3. To release from treatment those patients whose blood shows so much bactericidal power that no second attack need be feared—patients who would otherwise be forced to await the coming of the next attack at the hospital.

APPENDIX.

SUMMARY OF VIEWS EXPRESSED AT THE DISCUSSION ON SERUM DIAGNOSIS AT THE MEETING OF THE AMERICAN MEDICAL ASSOCIATION AT PHILADELPHIA, JUNE, 1897.

THE committee appointed by the Chairman of the Section on Practice of Medicine make the following report:

1. In selecting the material used in making the test the choice between: (*a*) serum, (*b*) dried blood, (*c*) fluid blood, and (*d*) blister fluid, will depend largely upon whether the object be scientific research, clinical diagnosis in hospital or private practice, or public laboratory diagnosis where the samples have to be sent some distance.

2. In spite of considerable variation in technique, there has been a remarkable uniformity in the results obtained by those taking part in the discussion, and their average of about ninety-five per cent of successes agrees with the general average of the cases, nearly four thousand, thus far recorded in medical literature.

3. Each of several methods of technique advocated may thus give good results in the hands of those thoroughly familiar with the details found necessary in each case and the sources of error to be avoided, success depending rather on being perfectly familiar with one method than on the particular one selected.

4. For routine diagnostic work even the very simplest methods may give good practical results, but for recording scientific observations those methods which are accurately quantitative should be selected. This is especially necessary in reporting exceptional cases at variance with the general results recorded or where the observations are made the basis of generalizations.

5. A complete reaction should comprise both characteristic clumping and total arrest of motion occurring within a definite time limit. For practical diagnostic work a dilution of 1:10, with a fifteen-minute time limit, is convenient. In any doubtful case the dilution should be carried as far as 1:50 or perhaps 1:60, and a reaction not obtainable at that point should not be regarded as perfectly conclusive. For these higher dilutions the time limit should be extended to two hours.

6. Intensity of reaction in a given serum should be estimated by determining the degree to which it may be diluted without losing its power of giving a decided reaction, both as to agglutination and loss of motion.

7. The intensity of reaction shown by the same serum is influenced by the age, condition, and virulence of the test culture and by the composition and reaction of the culture medium. For purposes of comparison the sensitiveness of the test culture should be taken into consideration.

8. The evidence so far recorded establishes that the reaction may be delayed or occasionally may not be obtained in cases of genuine typhoid infection; and also that it may be exceptionally present in non-typhoid cases, though not in an intense degree.

9. In investigating exceptional and contradictory results the following circumstances have to be considered: (a) The uncertainty of clinical diagnosis. (b) The absence of bacteriologic or other confirmatory methods of diagnosis during life, giving decisive negative results. (c) The possibility of overlooking typhoid infection even post mortem, in the absence of characteristic intestinal lesions where a very thorough bacteriologic examination has not been carried out.

10. The modifying influences mentioned above suffice to explain the divergencies existing in the reports of different observers. Without being absolutely infallible, the typhoid reaction appears to afford as accurate diagnostic results as can be obtained by any of the bacteriologic methods at our disposal for the diagnosis of other diseases. It must certainly be regarded as the most constant and reliable sign of typhoid fever, if not an absolute test.

N. B.: The above summary, while expressing the general consensus of opinion brought out during the discussion on serum diagnosis before the Section on Practice of Medicine of the

American Medical Association, does not claim to represent exactly the individual views of any one of those who took part.

W. H. WELCH,
WYATT JOHNSTON,
J. H. MUSSER,
R. C. CABOT,
S. S. KNEASS,
A. C. ABBOTT,
J. M. SWAN,

E. B. BLOCK,
H. M. BIGGS,
N. S. DAVIS, JR.,
M. W. RICHARDSON,
J. B. HERRICK,
A. R. GUERARD,
A. P. OHLMACHER,
Committee.

BIBLIOGRAPHY.

Aaser: Jour. Amer. Med. Assn., April 24, 1897.
Abbott: Med. News, May 8, 1897.
Achard: La Semaine méd., 1896, pp. 295, 301, 410, 480; 1897, p. 85. Discussion, December 10, 1896, Soc. des Hôpitaux.
Achard et Bensaude: 1. La Semaine méd., 1896, pp. 393, 480; 1897. 131. 2. Cholera, La Presse méd., Paris, September 26, 1896. Soc. Méd. des Hôp., April 23, 1897.
Achard et Lannelongue: Proteus Infections, La Semaine méd., 1896, p. 400.
Achard et Widal: Bull. et Mem. de la Soc. méd. des Hôp., August 6, 1896.
Aewenthal: Soc. Théra. de Moscow, March, 1897.
Aldridge: Lancet, No. 3899, May 21, 1898.
Alidieres: Thèse de Leipsic (Le Sero-diag. de Typh.).
Alpers and Murray: Amer. Med. Surg. Bull., N. Y., 1897, xi., p. 283.
Anderson: Lancet, May 27, 1897.
Antony et Ferré: Jour. de Méd. de Bordeaux, 1897, No. 30.
Appel and Thornbury: Jour. Amer. Med. Assn., Chicago, 1897, xxxiii., p. 241.
Archinard: Phil. Med. Jour., i., No. 6.
Archinard and Woolson: Note on Observ., N. Y. Med. News, 1898.
Arloing: Peripneumonia of Cattle, La Sem. méd., 1897, p. 38, and No. 20, 1898. Phil. Med. Jour., No. 25, 1898. Soc. de Méd. Lyon, February 15, 1897. Soc. de Biol., Paris, January 30, 1897.
Arnaud et Dassaud: Marseilles méd., September 15, 1897.
Arnaud et Eymeri: Limousin méd., November, 1897.
Auché et de Boncard: Jour. méd. de Bordeaux, August 30, 1896.
Barber: N. Y. Med. Jour., vol. lxvii., No. 16.
Bartlett: Yale Med. Jour., December, 1896, p. 78.
Bartoschevitch: Vratch, 1897, No. 2, p. 435.
Bebi: Gaz. degli Ospedali, September 20, 1896.
Beclère: Bull. Soc. méd. des Hôp., December 3, 1896.
Beco: Ann. Soc. méd.-chir. de Liège, 1896, xxxv., 303. Bull. de l'Acad. de Méd. Belgique, 1896, No. 11.
Behring und Nissen: Zeit. für Hyg., 1890, p. 424.
Berend: Orvosi Hetil., Buda-Pesth, 1897, xli., p. 52.
Bezançon: La Sem. méd., No. 22, 1898. La Presse méd., ii., 26, Paris, 1897.
Bezançon et Griffon: Soc. de Biol. Paris, June 5, 1897. Presse méd., July 17, 1897.
Biggs and Park: Amer. Jour. of Med. Sciences, February, 1897.
Blachstein: 1. Münch. med. Woch., November 3 and 10, 1896. 2. XIV. Cong. f. innere Med., Wiesbaden, April, 1896.
Blanquinque: Jour. des prat., February 13, 1897.
Block: Johns Hopkins Hosp. Bull., November, December, 1896; March, 1897; December 18, 1897. British Med. Assn., Montreal, September, 1897.
Blumenthal (Berlin): La Sem. méd., 152, 1897.
Bobi: Gaz. d. Osp., 1896, No. 113.

Bondet: Soc. des. Sc. méd., Lyon, February 15, 1897.
Bordet: 1. Annales de l'Inst. Pasteur, 1895, pp. 489, 492, 496; 1896, pp. 193, 211, 462. 2. Wien. med. Presse, 1896, p. 921. 3. La Presse méd., Paris, 1896, pp. 354, 374, 389.
Bormans: La Riforma med., Naples, 1896, vol. iv., pp. 579 and 590.
Bourget et Méry: La Semaine méd., No. 8.
Boyce: Lancet, February 13, 1897. Brit. Med. Assn., Montreal, September 2, 1897.
Bracken: N. Y. Med. Jour., April 14, 1897. Phil. Med. Jour., vol. 2, No. 8.
Brannan: N. Y. Med. Jour., March 27, 1897.
Breuer: Berl. klin. Woch., 1896, Nos. 47, 48.
Brill: N. Y. Med. Jour., lxvii., No. 2, March 2, 1897.
Brown: Brit. Med. Jour., March 12, 1898, No. 1941. Lancet, October 23, 1897.
Butters: Münch. med. Woch., 1897, 951.
Cabot: Boston Med. and Surg. Jour., February 4, 1897. Jour. Amer. Med. Assn., August 14, 1897.
Cahill: Med. News, April 10, 1897.
Cashin: Hog Cholera, Jour. Amer. Med. Assn., Chicago, 1897, p. 785.
Castaigne: Soc. de Biol., November 20, 1897. Gaz. hebdom., 1897, No. 55.
Catrin: La Sem. méd., 1896, pp. 410, 418. La Presse méd., October 17, 1896.
Chambrelant et St. Phillippe: Jour. de Méd. de Bordeaux, May 15, 1897. Soc. Gyn. et Obst. de Bordeaux, November 10, 1896.
Chantemesse: La Sem. méd., 1896, p. 301; 1897, p. 122.
Chantemesse et Ramond: Soc. méd. des Hôp., Paris, June 18, 1897.
Chantemesse et Widal: Annales de l'Inst. Pasteur, 1888, p. 54, 1892, p. 773.
Charriet et Apert: Compt. rendus de la Soc. de Biol., November 7, 1896. Presse méd., November 11, 1896.
Charrin: Compt. rend. de la Soc. de Biol., July 17, 1896.
Charrin et Ostrovsky: Compt. rend. de la Soc. de Biol, July 17, 1896.
Charrin et Roger: Compt. rend. de la Soc. de Biol., November 23, 1889, p. 667.
Christopher: Brit. Med. Jour., January 8, 1898.
Coleman: Lancet, January 16, 1897.
Colville and Donnan: Brit. Med. Jour., October 16, 1897.
Comba: Gazz. Osped. Clin., Milan, January 10, 1897.
Comba: La Riforma med., December 14, 1896, Nos. 288, 289.
Courmont: 1. Compt. rendus de la Soc. de Biol., July 25, 1896; February 5, 1897. 2. La Sem. méd., 1896, p. 294; and 1897, pp. 69, 105, 209. 3. Lyon méd., August 8, 1897. 4. La Presse méd., Paris, January 30, 1897.
Courmont et Martin: Lyon méd., 1897, No. 4.
Couture: Thèse de Paris, February 25, 1897 (La fièvre typh. chez l'enfant).
Craig: N. Y. Med. Jour., February 6, 1897.
Cruikshank: Can. Orat., Toronto, 1896, xxi., 371.
Dawson: Hog Cholera. N. Y. Med. Jour., February 20, 1897.
Debove: Soc. méd. des Hôp., May 7, 1897.
Delépine: 1. Med. Chron., Liverpool, October, 1896. Lancet, December 5 and 12, 1896; February 13, 1897. Brit. Med. Jour., February 27, April 17, 1897.
Dempsey: Lancet, January 16, 1897.
Derby: Arch. méd. Belges, 1896, p. 289.
Descazalo: Gaz. des Hôp., No. 111.
Deutsch: Gesellsch. der Aerzte, Buda-Pesth, February 20, 1897.
Devoto: Ref. in Maragliano's article, Chron. d. clin. med., Genoa, 1896, Anno iv., punt. 1.

Dieulafoy: Bull. méd., 1896, pp. 655, 933, Paris.
Dimoux-Dime: Thèse de Doct., Lyon, 1897, No. 5.
Dineur: Bull. Acad. méd. Belgique, xi., p. 705.
Du Cazal: Rev. gén. de Clin., Paris, 1896, x. 17. Rev. pract. de Méd., 1896, liii., p. 42.
Dumas: Thèse de Doct., Paris, 1896 (Sero-diag. de Widal).
Dungeon: Münch. med. Woch., June 29, 1897.
Dunlop: Brit. Med. Jour., December 11, 1897.
Dupaquier: New Orleans Med. and Surg. Jour., 1896, xlix., p. 458.
Durham: 1. Jour. of Path. and Bact., July, 1896. 2. Lancet, December 19, 1896; February 13, 1897, and No. 3881.
Elsberg: N. Y. Med. Record, April 10, 1897.
Engels: Centralblatt f. Bakt., January 30, 1897.
Eshner: Phil. Med. Jour., vol. i., No. 5, 1898.
Etienne: La Presse méd., Paris, 1896, p. 465.
Farchelt: Münch. med. Woch., 1897, 1061.
Feindel: Arch. gén. de Méd., October, 1897.
Ferrand et Theoari: 1. La Sem méd., 1897, p. 30. 2. Bull. Soc. méd. d. Hôp., January 22, February 4, 1897.
Fison: Brit. Med. Jour., July 31, 1897.
Förster: Zeit. f. Hyg., vol. xxiv., No. 3. Fortschritte der Med., June 1, 1897.
Fränkel, C.: 1. Hyg. Rundschau, 1896, No. 20. 2. Deut. med. Woch., January 14, April 15, 1897.
Fränkel, E.: Münch. med. Woch., 1897, No. 5.
Fränkel et Stadelmann: Soc. de Méd. int., Berlin, January 18, 1897.
Freyer: Brit. Med. Jour., August 7, 1897.
Fullerton: Lancet, May 1, 1897.
Gehrmann: Chicago Med. Record, May, 1897.
Gehrmann and Wynkoop: N. Y. Med. News, March, 1897.
Geraud and Remlinger: Bull. méd., Paris, April 21, 1897.
Gerloczy: Pester Med.-Chirurg. Presse, No. 23, 1897.
Gilbert et Fournier: Compt. rendus de la Soc. de Biol., December 4, 25, 1896. Bull. de l'Acad., October 20, 1896. Presse méd., January 16, 1897.
Gossage: West London Med. Jour., October, 1897.
Grandmaison: La Méd. moderne, December 12, 1896.
Greene: N. Y. Med. Record, November 14, December 5, 1896.
Griffith: Med. News, May 15, 1897.
Grüber: 1. Münch. med. Woch., March 3, 31, 1896; April 27, May 4, 1897. 2. Wien. klin. Woch., March 12, 1896. 3. XIV. Cong. für inn. Med., Wiesbaden, April, 1896.
Grünbaum: Lancet, September 19, December 19, 1896; February 13, 1897.
Guinon et Meunier: Soc. Méd. des Hôp., Paris, April 2, 1897.
Gwyn: Brit. Med. Jour., December 25, 1897. Johns Hopkins Hosp. Bull., vol. ix., No. 84.
Haedke: Deut. med. Woch., January 7, 1897.
Hager: Münch. med. Woch., 1897, p. 321.
Hammerschlag: Prag. med. Woch., July 29, 1897.
Hare: N. Y. Med. News, 1897, lxx., p. 490.
Haushalter: 1. La Sem. méd., 1896, p. 312. 2. Bull. méd., 1896, p. 769.
Hayem: La Sem. méd., 1897, p. 13. Bull. Soc. méd. des Hôp., January 8, 1897.
Hofmann: Centralbl. f. inn. Med., May 22, 1897.
Hua et Heberth: Normandie Méd., Rouen, September 1, 1896.
Hunt: Clin. Jour., London, ix., 345.
Issaeff: Annales de l'Inst. Pasteur, 1893, p. 269.
Issaeff und Ivanoff: Zeit. f. Hyg. u. Inf., 1894, p. 127.
Jackson: N. Y. Med. Jour., March 27, 1897.
Jemma: Centralbl. f. inn. Med., January 23, 1897.

Jez : Wien. med. Woch., January 16, 1897.
Johnston : N. Y. Med. Jour., October 31, 1896 ; January 16, June 5, 1897, p. 360, 1898.
Johnston and Hammond : Brit. Med. Jour., February 5, 1898.
Johnston and McTaggert : 1. Brit. Med. Jour., December 5, 1896 ; February 5, 1898. Montreal Med. Jour., March, 1897.
Johnston and Wolferstan : Brit. Med. Jour., February 5, 1898.
Josué et Clerc : Bull. Soc. Anat., fasc. xviii., 1896.
Jundell : Hygiea, Stockholm, 1896, p. 359.
Kerz : Brit. Med. Jour., December 11, 1897.
Kneas : Am. Med. Assn., Philadelphia, 1897.
Kolle : Centralbl. f. Bakt., vol. xix., Nos. 4 and 5. Deut. med. Woch.. 1897, No. 9 ; February 25, 1897.
Kolli : Vratch, 1896, Nos. 49 and 1381.
Kose : Cas. ces ck., 25, 25, 1897.
Kram : Wien. klin. Woch., 32, 1897.
Kraus : Wien. klin. Woch., April 30, 1897.
Kretz : Lancet, January 22, 1898.
Krüger : Gesellsch. f. inn. Med., Berlin, January 18, 1897. Deut. med. Woch., April 1, 1898.
Kuhnau : Ber. klin. Woch., No. 19, 1897.
Lambert : N. Y. Med. Jour., No. 9, 1898.
Landouzy et Griffon : Soc. de Biol., November 6, 1897.
Lannelongue et Achard : Compt. rendus de l'Acad. des Sc., October 5, 1896.
Lefevre : N. Y. Med. Jour., March 27, 1897 (see also Lambert, Loomis, Thatcher, Brill, etc., *ibidem* in discussion).
Lemoine : Bull. de la Soc. méd. des Hôp., July 30, 1896 ; May 7, 1897.
Lerch : Journ. of Amer. Med. Assn., February 26, 1898.
Le Sage : La Sem. méd., 383, 1897. Soc. Biol., October 16, 1897.
Leube : Münch. med. Woch., No. 8, 1898.
Levy und Bruns : Ber. klin. Woch., June 7, 1897.
Levy und Gissler : Münch. med. Woch., December 14, 21, 1897.
Lichtheim : Deut. med. Woch., 1896, No. 50.
Lindsay : Brit. Med. Jour., January 23, 1897.
Löffler und Abel : Centralbl. f. Bakter., xix., Nos. 2, 3.
Lowenthal : Deut. med. Woch., 35, September 16, 1897 ; Moscow, August 23, 1897.
Malvoz : Annales de l'Inst. Pasteur, July 25, 1897, tome xi., No. 7.
Maragliano : Cronaca d. clin. med. di Genoa, Anno 4, part 1.
Masius : La Sem. méd., 114, 1897.
Mathieu : Soc. méd. des Hôp., Paris, May 7, 1897.
McFarland : N. Y. Med. Jour., September 25, 1897.
McKensie : Canadian Pract., December, 1896.
McWeeney : Lancet, January 16, 1897 ; May 14, 1898.
McFadyean : Jour. Com. Path., December, 1896, p. 322.
Menétrier : 1. Bull. de la Soc. méd. des Hôp., July 30, December 4, 1896 ; 2. La Sem. méd., 1896, p. 497.
Metchnikoff : Annales de l'Inst. Pasteur, 1891, pp. 473 ; 1892, p. 296 ; 1894, pp. 187, 433, 714 ; 1895, p. 443.
Meunier et Guinon : La. Sem. méd., p. 121, 1897.
Miller : Johns Hopkins Hosp. Bull., vol. ix., p. 86. Chicago Med. Record, May, 1897.
Mills : La Sem. méd., 1897, 334 ; Moscow, August 23, 1897.
Mitchell : West. Med. Gaz., December, 1897.
Morillo : Thèse de doct., Paris, December, 1896 (La sero reaction).
Mosny et Issaeff, quoted by Vidal : Compt. rend. de la Soc. de Biol., December 25, 1896.
Mossé : Soc. méd. des Hôp., November 27, 1896. Compt. rend. Soc. de Biol., February 27, 1897. La Sem. méd., p. 76, 1897.
Mossé et Daunic : Soc. méd. des Hôp., Paris, March 5, 1897.

Moynier de Villepoix: Gazz. méd. de Picardie, Amiens, 1897, No. 13.
Murray: Lancet, May 29, 1897.
Musser and Swan: Brit. Med. Jour., December 18, 1897.
Nachod: Prag. med. Woch., July 29, 1897.
Nicolas: La Sem. méd., 1897, p. 37. Compt. rend. Soc. de Biol., July 25, 1896. Lyons méd., September 27, 1896. Soc. de Biol., Paris, December 5, 1896, iv., 106, 1897.
Nicolas et Charrin: La. Sem. méd., 1896, p. 496.
Nicolle et Halipré: Norm. méd., August 15, September 1, December 1 and 7, 1896.
Nicolle et Hébert: Norm. méd., Rouen, 1897, p. 95.
Ohlmacher: Cleveland Med. Mag., 1896, 1897, xii., p. 86.
Oordt (van): Münch. med. Woch., No. 3, 1897.
Osler: Edin. Med. Jour., November, 1897.
Pakes: Lancet, May 29, 1897.
Paltauf: Soc. Imp. Royale des Méd. de Vienna, May 28, 1897.
Pane: Centralbl. f. Bakt., xxi., p. 664.
Park: N. Y. Med. Record and Med. Jour., March 27, 1897; N. Y. Med. News, lxix., 559.
Patella: Atti e Rendiconti dell. Accad. med. chir. di Perugia, vol. ix., F. 2.
Paton: Brit. Med. Jour., December 11, 1897.
Pennato: Revista Veneta, January 31, 1897.
Périer: Jour. de Méd. de Paris, March 21, 1897.
Petermann: Vratch, No. 3, 1897.
Pfandler: Amer. Jour. Med. Sciences, vol. cxv., No. 4.
Pfeiffer: 1. Deut. med. Woch., November 29, 1894. 2. Zeit. f. Hyg., 1894, vols. 17, 18, 19, and 21. 3. Deut. med. Woch., 1896, Nos. 7, 12, 15. 4. XIV. Cong. f. inn. Med., Wiesbaden, April, 1896. 5. Centralbl. f. Bakt., 1896, vols. 19 and 20.
Pfeiffer und Issaeff: Deut. med. Woch., March 29, 1894.
Pfeiffer und Kolle: Deut. med. Woch., March 19, 1896, No. 12. Zeit. f. Hyg. u. Infect., 1896, vol. xxi., No. 2, p. 203. Centralbl. f. Bakt., January 30, 1897.
Pfühl: Centralbl. f. Bakt., No. 2, 1897.
Pick: Wien. klin. Woch., 1897, No. 4.
Prout: Am. Med. Surg. Bull., N. Y., 1897, xi., 186.
Pugliesi: La Riforma medica, Naples, 1896, vol. 12, p. 17.
Purjesz: Pest. med. Presse, Budapest, 1897, p. 159. Orvosi Hetilap., Budapest, 1897, No. 5.
Ransom: Lancet, January 16, 1897.
Reed: Johns Hopkins Med. Bull., March, 1897 (see also Thayer, *ibidem*).
Remlinger: Soc. Biol., July 10, 1897.
Renard: Thèse de Paris, 1897.
Rendu: Bull. de la Soc. méd. des Hôp., December 3, 1896.
Rénon: La Sem. méd., 1897, p. 38.
Richardson: Centralbl. f. Bakt., April 6, 1897.
Roberts: Lancet, February 13, 1897.
Robertson: Lancet, September 4, 1897.
Robin: Bull. méd., October 13, 1897.
Rochemont: Münch. med. Woch., No. 5, 1897.
Rodet: Compt. rend. de la Soc. de Biol., July 31, 1896; 366, 1897. La Sem. méd., No. 22, 1898.
Roger: Compt. rend. de la Soc. de Biol., July 4, 1896.
Roux: Acad. de Méd., July 13, 1897.
Rouget: Soc. méd. des Hôp., Paris, May 7, 1897.
Sabrazès: La Sem. méd., 1897, p. 13
Sabrazès et Hugon: Gaz. des Soc. méd. de Bordeaux, 1897, p. 28. Soc. méd. des Hôp., Paris, January 8, 1897.
Sabrazès and Rivière: Soc. de Biol., June 26, 1897.

Salimbeni: Annales de l'Inst. Pasteur, March 25, 1897.
Sanarelli: Annales de l'Inst. Pasteur, June, October 27, 1897. Policlinico, September 15, 1897.
Schäffer: Gaz. des Hôp., November 20, 1897.
Scheffer: Ber. klin. Woch., March 15, 1897.
Schwarz: Wien. Med.-Bl., 1897, p. 28.
Shattuck, G. B.: Med. News, May 8, 1897.
Shaw: Lancet, August 28, 1897.
Siegert: Münch. med. Woch., No. 10, 1897, p. 250.
Silvestrini: La Selt. Med., October 10, 1896.
Simon: Przeglad-Lekarska, February 15 and March 6, 1897.
Sinéff: Meditzinskae Obozrénié, Moscow, 1896, No. 22.
Siredey: Bull. de la Soc. méd. des Hôp., July 30, 1896.
Sklower: Thèse de Leipsic (Le sero-diag. de Typh.).
Skrjievan: Tougnorouss. neéditz. gazeta, Odessa, 1896, No. 51.
Soberheim: Hygien. Rundschau, December 1, 1896, No. 7. Zeitsch. f. Hyg., 1895, vol. xx.
Spiridonoff: Vratch, 1897, Nos. 22, 23.
Stadelmann: Soc. de Méd. Interne, Berlin, January 18 and April 11, 1897.
Starck: Münch. med. Woch., 1897, 374.
Stengel: N. Y. Med. Jour., vol. lxvii., No. 10.
Stern: 1. XIV. Cong. f. innere Med., Wiesbaden, April, 1896. 2. Centralbl. f. innere Med., December 5, 1896. 3. Berlin. klin. Woch., Nos. 11 and 12, 1897.
Stokes: N. Y. Med. Record, January 9, 1897.
Taty: Lyon méd., November 7, 1897.
Tew: Lancet, March 19, 1898.
Thatcher: N. Y. Med. Jour., March, 1897.
Thelliez: Thèse de doct., Paris, December, 1896 (Études sur le sero-diag.).
Theolen et Mills: La Clinique, Brussels, August 6 and September 3, 1896.
Therry and Mollison: Intercol. Med. Jour., Melbourne, Australia, 1897, p. 53.
Thiercelin: La. Sem. méd., 1896, p. 496.
Thiercelin et Lenoble: 1. Compt. rend. de la Soc. de Biol., December 5, 11, and 18, 1896. 2. La Presse méd., August 5, 1896.
Thiroloix: La Presse méd., November 4, 1896, No. 90.
Thoinot: La Semaine méd., 1896, p. 504.
Thoinot et Cavasse: Soc. méd. des Hôp., December 11, 1896.
Thomas: Med. News, April 3, 1897; lxxii., No. 18, 1898.
Thompson: Med. News, October 30, 1897.
Troisier et Sicard: Bull. de la Soc. méd. des Hôp., January 21, 1897.
Tschistovitch: Bolnitcharaya Geseta Bothnia, No. 51, 1897. Brit. Med. Jour., March 12, 1898.
Uhlenhuth: Deut. Militär-Zeit., Hft. 3, 1897.
Ullmann and Wöhnert: N. Y. Med. Jour., February 20, 1897.
Urban: Wien. med. Woch., August 7, 1897.
Vanlar et Beco: La Sem. méd., 1897, p. 15.
Vedel: La Sem. méd., 1896, p. 312.
Van der Velde: 1. La Sem. méd., 114, 384, 1897. 2. Bull. Acad. d. Méd. Belgique, No. 3, 1897.
Villiés et Battle: La Presse méd., October 10, 1896. Arch. gén., July, 1897.
Volykine: Vratch, December 25, 1897.
Volyntzew: Vratch, September 25, 1897.
Washburn: Jour. of Path. and Bact., April, 1895, vol. iii.
Wassermann: Zeit. f. Hyg. u. Infekt., 1896, p. 263.
Weaver: Chicago Med. Recorder, May, 1897.

Weinberg: La Presse méd., Paris, 1896, p. 682. Soc. de Biol., October 23, 1897.
Whittaker: Med. News, May 8, 1897.
Widal: 1. La Sem. méd., 1896, pp. 259, 269, 295, 303, 312, 488, 497; 1897, No. 69; 1898, No. 19. 2. Jour. de Méd., July 25, 1896; 1897, pp. 122, 295, 332, 334, 384. 3. La Presse méd., Paris, July 29, 1896. 4. Lancet, November 14, 1896.
Widal et Noblecourt: La Sem. méd., p. 285, 1897.
Widal et Sicard: 1. La Sem. méd., 1896, pp. 393, 410, 418, 488, 497, 504; 1897, pp. 14, 21, 38, 69; 1898, No. 19. 2. La Presse méd., October 4, 1896. 3. Bull. de l'Acad. de Méd., 1896, p. 347. 4. Compt. rendus de la Soc. de Biol., December 25, 1896. 5. Annales de l'Inst. Pasteur, No. 5, 1897.
Wilson: Phil. Med. Jour., vol. i., No. 13.
Wilson and Westbrooke: Brit. Med. Jour., December 18, 1897.
Wright: Brit. Med. Jour., January 16, 1897, and February 5, 1898, p. 355.
Wright and Semple: Brit. Med. Jour., January 30; May 15, 1897.
Wright and Smith: Lancet, 1897, No. 10. Brit. Med. Jour., April 10, 1897.
Zabolotny: Deut. med. Woch., June 10, 1897.
Ziemke: Deut. med. Woch., April 8, 1897.

INDEX.

ABBOTT, 28
Achard, 9, 41, 49
Agglutination, 4, 13
Agglutinative centres, 14
    substance, origin and nature of, 56-60
Anthrax, 139
Appearance of the reaction, time of, 66
Auto-serum reaction, 55

BECO and Vanlair, 24
Beusaude, 22
Biggs and Park, 18, 27, 42, 46
Bile, clumping power of, 51
Blister fluid, 18
Block, 27, 45
Body fluid, power of, 48
Bordet, 8, 30
Bormans, 49
Bubonic plague, 124

CENTRIFUGALIZING the blood, 18
Chain-formation, 20
Chantemesse, 9
Charron, 6
Chemical reagents, clumping with, 55
Cholera, 120
Coleman, 17
Collection of blood, 11
Colon-bacillus infections, 127
Contamination, 11
Courmont, 9, 51
Cultures, difference of races, 40
    effect of environment, 43

DA COSTA, 28
Dead bacilli, use of, 30

Delepine, 17, 32
Diagnosis, Widal's test in, 2
Dieulafoy, 9
Dilution, 12, 15
Diphtheria, 135
Disappearance of the reaction, 73
Doubtful epidemics, 105
Dried blood, use of, 26
Durham, 8, 40

FEVER, relation to reaction, 78
Filtered cultures, reactions with, 54
Fœtus, transmission to, 52
Förster, 42
Fraenkel, 16, 41

GEHRMANN and Wynkoop, 28
Glanders, 117
Greene, 20, 22
Gruber, 8, 16
Grünbaum, 16

HAEDKE, 16, 20, 23
Halpré, 9
History of serum diagnosis, 6-10
Hog cholera, 7, 137

INTERMITTENCE of reaction, 78
Isaeff, 7

JOHNSTON, 9, 10, 28, 45

KOLLE, 16
Kühnau, 16

LATE appearance of reaction, 72
    disappearance of reaction, 75
    reactions, theory of, 77
Leprosy, 134

MACROSCOPIC TESTS, 21
Malta fever, 118
McFarland, 20
McWeeney, 17, 19, 20
Measurement of clumping power, 31
Menetrier, 9
Metchnikoff, 7
Microscopic methods, 10-20
Milk, clumping power of, 49
Miller, 9
Mills, 42

NASCENT CULTURES, use of, 21
Nicolle, 9

PASTEUR'S PIPETTE, use of, 18
Peripneumonia of cattle, 136
Pfeiffer, 8
Pfeiffer's phenomenon, 8
Pfuhl, 17, 27
Pick, 27
Pictou cattle disease, 137
Pneumococcus, 7
    infections, 125
Proteus infections, 137
Psittacosis, 139
Pugliesi, 18
Pus, clumping power of, 48
Pyocyaneus, 6

RELAPSING FEVER, 141
Retrospective diagnosis, 105
Richardson, 42
Rochemont, 16
Roger, 6

SABRAZÉS and Hugon, 23
Scheffer, 16
Sedimentation tubes, 25
Siredy, 9
Serous effusion, clumping power of, 50

Sero-prognosis, 113
Sources of error, 15, 21, 23
Staining clumps, 20
Staphylococcus infections, 139
Stern, 16, 18, 23, 32, 46
Stokes, 20
Stools, clumping power of, 49
Suspensions, use of, 19
Streptococcus infections, 132

TEARS, clumping power of, 50
Technique, 10, 29
Tetanus, 136
Time limit, 15
Transmission through placenta, 52
Tropics, Widal's test in, 1-4
Tuberculosis, 132
Typhoid, abortive, 106
    and allied infections, 110
    and Mediterranean fever, 112
    and mountain fever, 112
    cholecystitis, 108
    general statistics, 61-63
    in infancy and old age, 106
    melancholic forms, 109
    meningitis, 109
    with double infection, 108
    without fever, 106
    without intestinal lesions, 107

URINE, clumping power of, 49

VACCINATION, effect of, 78
Value of serum diagnosis, 1, 79, 89

WHOLE BLOOD, use of, 17
Widal, 15, 17, 32, 40
Wilson and Westbrook, 17, 28
Wright, 10, 24, 31

YELLOW FEVER, 2, 119

www.ingramcontent.com/pod-product-compliance
Lightning Source LLC
Chambersburg PA
CBHW030248170426
43202CB00009B/670